Lecture Notes in Computer Science   7980

Commenced Publication in 1973
Founding and Former Series Editors:
Gerhard Goos, Juris Hartmanis, and Jan van Leeuwen

Abdelkader Hameurlain   Josef Küng
Roland Wagner (Eds.)

Transactions on Large-Scale Data- and Knowledge-Centered Systems IX

 Springer

Editors-in-Chief

Abdelkader Hameurlain
Paul Sabatier University, IRIT
118, route de Narbonne, 31062 Toulouse Cedex, France
E-mail: hameur@irit.fr

Josef Küng
Roland Wagner
University of Linz, FAW
Altenbergerstraße 69, 4040 Linz, Austria
E-mail: {jkueng, rrwagner}@faw.at

ISSN 0302-9743 (LNCS)   e-ISSN 1611-3349 (LNCS)
ISSN 1869-1994 (TLDKS)
ISBN 978-3-642-40068-1   e-ISBN 978-3-642-40069-8
DOI 10.1007/978-3-642-40069-8
Springer Heidelberg Dordrecht London New York

Library of Congress Control Number: 2013944558

CR Subject Classification (1998): H.2.8, H.2, C.2, F.2, I.2.6, H.3, J.1

Typesetting: Camera-ready by author, data conversion by Scientific Publishing Services, Chennai, India

Printed on acid-free paper

Springer is part of Springer Science+Business Media (www.springer.com)

Preface

The LNCS journal Transactions on Large-Scale Data- and Knowledge-Centered Systems focuses on data management, knowledge discovery, and knowledge processing, which are core and hot topics in computer science.

This volume is the third so-called regular volume of the TLDKS journal. It contains, after reviewing, a selection of 5 contributions from 20 submitted papers in response to the call for papers for this regular volume. One paper has been selected from the 5th International Conference on Data Management in Cloud, Grid and Peer-to-Peer Systems (Globe 2012), which was held during September 5–6, 2012, in Vienna, Austria. The authors were invited to submit an extended version for a new round of reviewing.

The content of this volume covers a wide range of different and hot topics in the field of data and knowledge management, mainly: top-k query processing in P2P systems, self-stabilizing consensus average algorithms in distributed sensor networks, recoverable encryption schemes, XML data in a multi-system environment, and pairwise similarity for cluster ensemble problems.

We would like to express our thanks to the external reviewers and editorial board for thoroughly refereeing the submitted papers and ensuring the high quality of this volume. Special thanks go to Gabriela Wagner for her availability and her valuable work in the realization of this TLDKS volume.

May 2013

Abdelkader Hameurlain
Josef Küng
Roland Wagner

Editorial Board

Table of Contents

As-Soon-As-Possible Top-k Query Processing in P2P Systems

William Kokou Dédzoé[1], Philippe Lamarre[2], Reza Akbarinia[3], and Patrick Valduriez[3]

[1] INRIA Rennes, France
William.Dedzoe@inria.fr
[2] INSA de Lyon, France
Philippe.Lamarre@insa-lyon.fr
[3] INRIA and LIRMM, Montpellier, France
{Reza.Akbarinia,Patrick.Valduriez}@inria.fr

Abstract. Top-k query processing techniques provide two main advantages for unstructured peer-to-peer (P2P) systems. First they avoid overwhelming users with too many results. Second they reduce significantly network resources consumption. However, existing approaches suffer from long waiting times. This is because top-k results are returned only when all queried peers have finished processing the query. As a result, query response time is dominated by the slowest queried peer. In this paper, we address this users' waiting time problem. For this, we revisit top-k query processing in P2P systems by introducing two novel notions in addition to response time: the *stabilization time* and the *cumulative quality gap*. Using these notions, we formally define the as-soon-as-possible (ASAP) top-k processing problem. Then, we propose a family of algorithms called ASAP to deal with this problem. We validate our solution through implementation and extensive experimentation. The results show that ASAP significantly outperforms baseline algorithms by returning final top-k result to users in much better times.

1 Introduction

Unstructured *Peer-to-Peer* (P2P) systems have gained great popularity in recent years and have been used by millions of users for sharing resources and content over the Internet [4,30,25]. In these systems, there is neither a centralized directory nor any control over the network topology or resource placement. Because of few topological constraints, they require little maintenance in highly dynamic environnements [26]. However, executing queries over unstructured P2P systems typically by flooding may incur high network traffic and produce lots of query results.

To reduce network traffic and avoid overwhelming users with high numbers of query results, complex query processing techniques based on top-k answers have been proposed e.g. in [2]. With a top-k query, the user specifies a number k of the most relevant answers to be returned by the system. The quality (i.e. score of relevance) of the answers to the query is determined by user-specified scoring

A. Hameurlain et al. (Eds.): TLDKS IX, LNCS 7980, pp. 1–27, 2013.

functions [9,18]. Despite the fact that these top-k query processing solutions e.g. [2] reduce network traffic, they may significantly delay the answers to users. This is because top-k results are returned to the user only when all queried peers have finished processing the query. Thus, query response time is dominated by the slowest queried peer, which makes users suffer from long waiting times. Therefore, these solutions are not suitable for emerging applications such as P2P data sharing for online communities, which may have high numbers of autonomous data sources with various access performance. Most of the previous work on top-k processing has focused on efficiently computing the exact or approximate result sets and reducing network traffic [6,17,34,32,2].

A naive solution to reduce users' waiting time is to have each peer return its top-k results directly to the query originator as soon as it has done executing the query. However, this significantly increases network traffic and may cause a bottleneck at the query originator when returning high numbers of results. In this paper, we aim at reducing users' waiting time by returning high quality intermediate results, while avoiding high network traffic. The intermediate results are the results of peers which have already processed locally their query. Providing intermediate results to users is quite challenging because a naive solution may saturate users with results of low quality, and incur significant network traffic which in turn may increase query response time.

In this paper, our objective is to return high quality results to users as soon as possible. For this, we revisit top-k query processing in P2P systems by introducing two notions to complement response time: *stabilization time* and *cumulative quality gap*. The stabilization time is the time needed to obtain the final top-k result set, which may be much lower than the response time (when it is sure that there is no other top-k result). The quality gap of the top-k intermediate result set is the quality that remains to be the final top-k result set. The cumulative quality gap is the sum of the quality gaps of all top-k intermediate result sets during query execution. Using these notions, we formally define the as-soon-as-possible (ASAP) top-k processing problem. Then, we propose a family of algorithms called ASAP to deal with this problem.

This paper is an extended version of [12] with the following added value. First, in Section 6 we propose a solution to deal with node failures (or departures) which may decrease the quality and accuracy of top-k results. In Section 7, we propose two techniques to compute "probabilistic guarantees" for the users showing for example the probability that current intermediate top-k results are the true top-k results (i.e. confidence of current top-k result). Section 8.2 shows experimentally the effectiveness of our solution for computing "probabilistic guarantees". Finally, we study experimentally the impact of data distribution on our algorithms (Section 8.2).

2   System Model

In this section, we first present a general model of unstructured P2P systems which is needed for describing our solution. Then, we provide a model and definitions for top-k queries.

2.1 Unstructured P2P Model

We model an unstructured P2P network of n peers as an undirected graph $G = (P, E)$, where $P = \{p_0, p_1, \cdots, p_{n-1}\}$ is the set of peers and E the set of connections between the peers. For $p_i, p_j \in P, (p_i, p_j) \in E$ denotes that p_i and p_j are neighbours. We also denote by $N(p_i)$, the set of peers to which p_i is directly connected, so $N(p_i) = \{p_j | (p_i, p_j) \in E\}$. The value $\|N(p_i)\|$ is called the degree of p_i. The average degree of peers in G is called the *average degree* of G and is denoted by φ. The *r-neighborhood* $N^r(p)$ $(r \in \mathbb{N})$ of a peer $p \in P$ is defined as the set of peers which are at most r hops away from peer p, so

$$N^r(p) = \begin{vmatrix} \{p\} & \text{if } r = 0 \\ \{p\} \bigcup_{p' \in N(p)} N^{r-1}(p') & \text{if } r \geq 1 \end{vmatrix}$$

Each peer $p \in P$ holds and maintains a set $D(p)$ of data items such as images, documents or relational data (i.e. tuples). We denote by $D^r(p)(r \in \mathbb{N})$, the set of all data items which are in $N^r(p)$, so

$$D^r(p) = \bigcup_{p' \in N^r(p)} D(p')$$

In our model, the query is forwarded from the query originator to its neighbours until the Time-To-Live value of the query decreases to 0 or the current peer has no peer to forward the query. So the query processing flow can be represented as a tree, which is called the query forwarding tree. When a peer $p_0 \in P$ issues query q to peers in its *r-neighborhood*, the results of these peers are bubbled up using query q's forwarding tree with root p_0 including all the peers belonging to $N^r(p_0)$. The set of children of a peer $p \in N^r(p_0)$ in query q's forwarding tree is denoted by $\psi(p, q)$.

2.2 Top-k Queries

We characterize each top-k query q by a tuple $< qid, c, ttl, k, f, p_0 >$ such that qid is the query identifier, c is the query itself (e.g. SQL query), $ttl \in \mathbb{N}$ (Time-To-Live) is the maximum hop distance set by the user, $k \in \mathbb{N}^*$ is the number of results requested by the user $f : \mathcal{D} \times \mathcal{Q} \to [0,1]$ is a scoring function that denotes the score of relevance (i.e. the quality) of a given data item with respect to a given query and $p_0 \in P$ the originator of query q, where $\mathcal{D}$ is the set of data items and $\mathcal{Q}$ the set of queries.

A top-k result set of a given query q is the k top results among data items owned by all peers that receive q. Formally we define this as follows.

Definition 1 *Top-k Result Set*. *Given a top-k query q, let $D' = D^{q.ttl}(q.p_0)$. The top-k result set of q, denoted by $Top^k(D', q)$, is a sorted set on the score (in decreasing order) such that:*

1. *$Top^k(D', q) \subseteq D'$;*
2. *If $\|D'\| < q.k$, $Top^k(D', q) = D'$, otherwise $\|Top^k(D', q)\| = q.k$;*
3. *$\forall d \in Top^k(D', q), \forall d' \in D' \setminus Top^k(D', q),   q.f(d, q.c) \geq q.f(d', q.c)$*

Definition 2 *Result's Rank*. *Given a top-k Result set I, we define the rank of result $d \in I$, denoted by $rank(d, I)$, as the position of d in the set I.*

Note that the rank of a given top-k item is in the interval $[1; k]$.

In large unstructured P2P systems, peers have different processing capabilities and store different volumes of data. In addition, peers are autonomous in allocating the resources to process a given query. Thus, some peers may process more quickly a given query than others. Intuitively, the top-k intermediate result set for a given peer is the k best results of both the results the peer received so far from its children and its local results (if any). Formally, we define this as follows.

Definition 3 *Top-k Intermediate Result Set*. *Given a top-k query q, and $p \in N^{q.ttl}(q.p_0)$. Let D_1 be the result set of q received so far by p from peers in $\psi(p, q)$ and $D_2 = D_1 \cup D(p)$. The top-k intermediate result set of q at peer p, denoted by $I_q(p)$, is such that:*

$$I_q(p) = \begin{vmatrix} Top^k(D_2, q) \text{ if } p \text{ has already processed } q \\ \\ Top^k(D_1, q) \text{ otherwise} \end{vmatrix}$$

3   Problem Definition

Let us first give our assumptions regarding schema management and the unstructured P2P architecture. We assume that peers are able to express queries over their own schema without relying on a centralized global schema as in data integration systems [28]. Several solutions have been proposed to support decentralized schema mapping. However, this issue is out of scope of this paper and we assume it is provided using one of the existing techniques, *e.g.* [23], [28] and [1]. We also assume that all peers in the system are trusted and cooperative. In the following, we first give some definitions which are useful to define the problem we focus and formally state the problem.

3.1   Foundations

To process a top-k query in P2P systems, an ASAP top-k algorithm provides intermediate results to users as soon as peers process the query locally. This allows

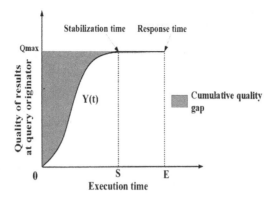

Fig. 1. Quality of top-k results at the query originator wrt. Execution time.

users to progressively see the evolution of their query execution by receiving intermediate results for their queries. Note that at some point of query execution, the top-k intermediate results received by a peer may not change any more, until the end of the query execution. We denote this point as the **stabilization time** (see Figure 1).

Recall that the main goal of ASAP top-k query processing is to return high-quality results to user as soon as possible. To reflect this, we introduce the *quality evolution* concept. Given a top-k query q, we define the quality evolution $Y(t)$ of q at time t as the sum of scores of q's intermediate top-k results at t and at q's originator. Figure 1 shows the quality evolution of intermediate top-k results obtained at the query originator during a given query execution. To be independent of the scoring values —which can be different from one query to another—, we normalize the quality evolution of a query. With this in mind, we divide the quality evolution of a given query by the sum of scores of the final top-k results of that query. Thus, the quality evolution values are in the interval $[0, 1]$ and the quality of the top-k final results is equal to 1. Note that we do not use the proportion of the final top-k results in intermediate top-k results (i.e precision) to characterize ASAP algorithm because this metric does not express the fact of returning the high quality results as soon as possible to users.

The quality evolution of intermediate top-k results at the query originator increases as peers answer the query. To reflect this, we introduce the *cumulative quality gap*, which is defined as the sum of the quality difference between intermediate top-k result sets received until the stabilization time and the final top-k result set. We formalize this in Definition 4.

Definition 4 *Cumulative Quality Gap.* *Given a top-k query q, let $Y(t)$ be the quality evolution of q at time t at q originator, and S be the stabilization time of q. The cumulative quality gap of the query q, denoted by C_{qg} is:*

$$C_{qg} = \int_{0}^{S} (1 - Y(t))\, \mathrm{d}t = S - \int_{0}^{S} Y(t)\, \mathrm{d}t \qquad (1)$$

3.2   Problem Statement

Formally, we define the ASAP top-k query processing problem as follows. Given a top-k query q, let S be the stabilization time of q and C_{qg} be the cumulative quality gap of q. The problem is to minimize C_{qg} and S while avoiding high communication cost.

4   ASAP Top-k Query Processing Overview

ASAP query processing proceeds in two main phases. The first phase is the query forwarding and local execution of the query. The second phase is the bubbling up of the peers' results for the query along the query forwarding tree.

4.1   Query Forwarding and Local Execution

Query processing starts at the query originator, i.e. the peer at which a user issues a top-k query q. The query originator performs some initialization. First, it sets ttl which is either user-specified (or default). Second, it creates a unique identifier qid for q which is useful to distinguish between new queries and those received before. Then, q is included in a message that is broadcast by the query originator to its reachable neighbors. **Algorithm 1** shows the pseudo-code of query forwarding. Each peer that receives the message including q checks qid (see line 2, Algorithm 1). If it is the first time the peer has received q, it saves the query (i.e. saves the query in the list of seen queries and the address of the sender as its parent) and decreases the query ttl by 1 (see lines 3-4, Algorithm 1). If the ttl is greater than 0, then the peer sends the query message to all neighbors except its parent (see lines 5-7, Algorithm 1). Then, it executes q locally. If q has been already received, then if the old ttl is smaller than the new ttl, the peer proceeds as where q is received for the first time but without executing q locally (see lines 10-18, Algorithm 1), else the peer sends a duplicate message to the peer from which it has received q.

4.2   Bubbling Up Results

Recall that, when a peer submits a top-k query q, the local results of the peers who have received q are bubbled (i.e returned upwards) up to the query originator using query q's forwarding tree. In ASAP, a peer's decision to send intermediate results to its parent is based on the improvement impact computed by using the ratio of its current top-k intermediate result set over the top-k intermediate result set which it has sent so far to its parent. This improvement impact can be computed in two ways: by using the score or rank of top-k results in the result set. Therefore, we introduce two types of improvement impact: *score-based improvement impact* and *rank-based improvement impact*.

Intuitively, the score-based improvement impact at a given peer for a given top-k query is the gain of score of that peer's current top-k intermediate set compared to the top-k intermediate set it has sent so far.

Algorithm 1. *receive_Query(msg)*

---

input : *msg*, a query message.
1 **begin**
2   **if** *(!already_Received(msg.getID()))* **then**
3   *memorize(msg);*
4   *msg.decreaseTTL();*
5   **if** *(msg.getTTL() > 0)* **then**
6   *forwardToNeighbors(msg);*
7   **end**
8   *executeLocally(msg.getQuery());*
9   **else**
10   *qid = msg.getID();*
11   *oldMsg = SeenQuery(qid).;*
12   **if** *(msg.getTTL() > oldMsg.TTL())* **then**
13   *memorize(msg);*
14   *msg.decreaseTTL();*
15   **if** *(msg.getTTL() > 0)* **then**
16   *forwardToNeighbors(msg);*
17   **end**
18   *sendDuplicateSignal(qid, oldMsg.getSender());*
19   **else**
20   *sendDuplicateSignal(qid, msg.getSender());*
21   **end**
22   **end**
23 **end**

Definition 5 *Score-Based Improvement Impact.* *Given a top-k query q, and peer $p \in N^{q.ttl}(q.p_0)$, let T_{cur} be the current top-k intermediate set of q at p and T_{old} be the top-k intermediate set of q sent so far by p. The score-based improvement impact of q at peer p, denoted by $IScore(T_{cur}, T_{old})$ is computed as*

$$IScore(T_{cur}, T_{old}) = \frac{\sum_{d \in T_{cur}} q.f(d, q.c) - \sum_{d' \in T_{old}} q.f(d', q.c)}{k} \qquad (2)$$

Note that in Formula 2, we divide by k instead of $\|T_{cur} - T_{old}\|$ because we do not want that $IScore(T_{cur}, T_{old})$ be an average which would not be very sensitive to the values of scores. The score-based improvement impact values are in the interval $[0, 1]$.

Intuitively, the rank-based improvement impact at a given peer for a given top-k query is the loss of rank of results in the top-k intermediate result set sent so far by that peer due to the arrival of new intermediate results.

Definition 6 *Rank-Based Improvement Impact.* *Given a top-k query q and peer $p \in N^{q.ttl}(q.p_0)$, let T_{cur} be the current top-k intermediate result set of q at p and T_{old} be the top-k intermediate result set of q sent so far by p. The rank-based improvement impact of q at peer p, denoted by $IRank(T_{cur}, T_{old})$ is computed as*

$$IRank(T_{cur}, T_{old}) = \frac{\sum_{d \in T_{cur} \setminus T_{old}} (k - rank(d, T_{cur}) + 1)}{\frac{k * (k + 1)}{2}} \qquad (3)$$

Note that in Formula 3, we divide by $\frac{k*(k+1)}{2}$ which is the sum of ranks of a set containing k items. The rank-based improvement impact values are in the interval $[0, 1]$.

Notice also that, in order to minimize network traffic, ASAP does not bubble up the results (which could be large), but only their scores and addresses. A score-list is simply a list of k pairs (ad, s), such that ad is the address of the peer owning the data item and s its score.

A simple way to decide when peer must bubble up newly received intermediate results to its parent is to set a minimum value (threshold) that must reach its improvement impact. This value is set initially by the application and it is the same for all peers in the system. Note also that this threshold does not change during the execution of the query. Using both types of improvement impact we have introduced, we have two types of static threshold-based approaches. The first approach uses the score-based improvement impact and the second one the rank-based improvement impact.

A generic algorithm for our static threshold-based approaches is given in **Algorithm 2**. In these approaches, each peer maintains for each query a set T_{old} of top-k intermediate results sent so far to its parent and a set T_{cur} of current top-k intermediate results. When a peer receives a new result set N from its children (or its own result set after local processing of a query), it first updates the set T_{cur} with results in N (see line 2, Algorithm 2). Then, it computes the improvement impact imp of T_{cur} compared to T_{old} (line 3, Algorithm 2). If imp is greater than or equal to the defined threshold $delta$ or if there are no more children' results to wait for, the peer sends the set $T_{tosend} = T_{cur} \setminus T_{old}$ to its parent and subsequently sets T_{curr} to T_{old} (see lines 4-7, Algorithm 2).

5   Dynamic Threshold-Based Approaches for Bubbling Up Results

Although the static threshold-based approaches are interesting to provide results quickly to user, they may be blocking if results having higher scores are bubbled up before those of lower score. In other words, sending higher score's results will induce a decrease of improvement impact of the following results. This is because the improvement impact considers the top-k intermediate results sent so far by the peer. Thus, results of low scores even if they are in the final top-k results may be returned at the end of the query execution. To deal with this problem, an interesting way would be to have a dynamic threshold, i.e. a threshold that decreases as the query execution progresses. However, this would require finding the right parameter on which the threshold depends. We have identified two possible solutions for the dynamic threshold. The first one is to use an estimation of the query execution time. However, estimating the query execution time in large P2P system is very difficult because it depends on network dynamics, such as connectivity, density, medium access contention, etc., and the slowest queried peer. The second, more practical, solution is to use the peer's result set coverage, i.e for each peer the proportion of peers in its sub-tree including itself (i.e. all

---

Algorithm 2. $Streat(k, T_{cur}, T_{old}, N, delta, Func)$

---

    input : k, number of results; T_{cur}, current top-k; T_{old}, top-k sent so far; N, new
               result set; $delta$, impact threshold; $Func$, type of improvement impact.

1 **begin**
2 | $T_{cur} = mergingSort\_Topk(k, T_{cur}, N)$;
3 | $imp = Func(T_{cur}, T_{old})$;
4 | **if** $((imp \geq delta)$ **or** $all\_Results())$ **then**
5 | | $T_{tosend} = T_{cur} \setminus T_{old}$;
6 | | $send\_Parent(T_{tosend}, all\_Results())$;
7 | | $T_{old} = T_{cur}$;
8 | **end**
9 **end**

---

its descendants and itself) which have already processed the query to decrease the threshold.

5.1 Peer's Local Result Set Coverage

Definition 7 *Peer's Local Result Set Coverage.* *Given a top-k query, and $p \in N^{q.ttl}(q.p_0)$, let $\mathcal{A}$ be the set of peers in the sub-tree whose root is p in the query q's forwarding tree. Let $\mathcal{E}$ be the set of peers in $\mathcal{A}$ which have already processed q locally. The local result set coverage of peer p for q, denoted by $Cov(\mathcal{E}, \mathcal{A})$, is computed using the following equation:*

$$Cov(\mathcal{E}, \mathcal{A}) = \frac{\|\mathcal{E}\|}{\|\mathcal{A}\|}$$

Peer's local result set coverage values are in the interval $[0, 1]$.

Note that is very difficult to have the exact value of a peer's local result set coverage without inducing an additional number of messages in the network. This is because each peer must send a message to its parent each time its local coverage result set value changes. Thus, when a peer at hop m from query originator updates its local result coverage, m messages will be sent over the network. To deal with this problem, an interesting solution is to have an estimation of this value instead of the exact value.

The estimation of peer's local result set coverage can be done using two different strategies: optimistic and pessimistic. In the optimistic strategy, each peer computes the initial value of its local result set coverage based only on its children nodes. This value is then updated progressively as the peers in its sub-tree bubble up their results. Indeed, each peer includes in each response message sent to its parent the number of peers in its sub-tree (including itself) which have already processed the query locally and the total number of peers in its sub-tree including itself. This couple of values is used in turn by its parent to estimate its local result set coverage. Contrary to the optimistic strategy, in the pessimistic strategy, the local result set coverage estimation is computed at the beginning by each peer based on the Time-To-Live received with the query and the average degree of peers in the system. As in the case of the optimistic strategy, this value is updated progressively as the peers in its sub-tree bubble up their results.

In our dynamic threshold-based approaches, we estimate a peer's local result set coverage using the pessimistic strategy because the estimation value is more stable than with the optimistic strategy. Now, let us give more details about how a peer's local result set coverage pessimistic estimation strategy is done.

5.2   Peer's Local Result Set Coverage Pessimistic Estimation

In order to estimate its local result set coverage, each peer p_i maintains for each top-k query q and for each child p_j a set $\mathcal{C}_1$ of pairs (p_j, a) where $a \in \mathbb{N}$ is the number of peers in the sub-tree of peer p_j including p_j itself. p_i maintains also a set $\mathcal{C}_2$ of pairs (p_j, e) where $e \in \mathbb{N}$ is the total number of peers in the sub-tree of peer p_j including p_j itself which have already processed locally q. Now let ttl' be the time-to-live with which p_i received query q and φ be the average degree of peers in the system. At the beginning of query processing, for all children of p_i, $e = 0$ and $a = \sum_{u=0}^{ttl'-2} \varphi^u$. During query processing, when a child p_j in $\psi(p_i, q)$ wants to send results to p_i, it inserts in the answer message its couple of values (e, a). Once p_i receives this message, it unpacks the message, gets these values (i.e. e and a) and updates the sets $\mathcal{C}_1$ and $\mathcal{C}_2$. The local result set coverage of peer p_i for the query q is then estimated using Formula 4.

$$\widetilde{Cov}(\mathcal{C}_1, \mathcal{C}_2) = \frac{\sum\limits_{(p_j,e)\in\mathcal{C}_1} e}{\sum\limits_{(p_j,a)\in\mathcal{C}_2} a} \tag{4}$$

Note that peer's local result set coverage estimation values are in the interval $[0, 1]$.

5.3   Dynamic Threshold Function

In the dynamic threshold approaches, the improvement impact threshold used by a peer at a given time t of the query execution depends on its local result set coverage at that time. This improvement impact threshold decreases as the local result set coverage increases. To decrease the improvement impact threshold used by a peer as the local result set coverage increases, we use a linear function that allows peers to set their improvement impact threshold for a given local result set coverage. Now let us define formally the threshold function.

Definition 8 Dynamic Threshold Function. *Given a top-k query q and $p \in N^{q.ttl}(q.p_0)$, the improvement impact threshold used by p during q's execution, is a monotonically decreasing function H such that:*

$$H : \left| \begin{array}{l} [0, 1] \to [0, 1] \\ \\ x \quad \mapsto -\alpha * x + \alpha \end{array} \right. \tag{5}$$

Algorithm 3. $Dtreat(k, T_{cur}, T_{old}, N, Func, cov, \bar{cov}, H)$

---

  input : k; T_{cur}; T_{old}; N; $Func$; cov, current local result set coverage; $\bar{cov}$, result set coverage threshold; H, a dynamic threshold function.

1 **begin**
2    $T_{cur} = mergingSort\_Topk(k, T_{cur}, N)$;
3   **if** $(cov > \bar{cov})$ **then**
4      $delta = H(cov)$;
5      $imp = Func(T_{cur}, T_{old})$;
6   **if** $((imp \geq delta)$ **or** $all\_Results())$ **then**
7        $T_{tosend} = T_{cur} \setminus T_{old}$;
8        $send\_Parent(T_{tosend}, all\_Results())$;
9        $T_{old} = T_{cur}$;
10   **end**
11   **end**
12 **end**

---

with $\alpha \in [0, 1[$. Notice that x is a peer's result set coverage at given time and α the initial improvement impact threshold (i.e. $H(0) = \alpha$).

5.4   Reducing Communication Cost

Using a rank-based improvement impact has the drawback of not reducing as much as possible network traffic. This is because the rank-based improvement impact value is equal to 1 (the maximum value it can reach) when a peer receives the first result set (from one of its children or after local processing of a query). Thus, each peer always sends a message over the network when it receives the first result set containing k results. To deal with this problem and thus reduce communication cost, we use peers' result sets coverage to prevent them to send a message when they receive their first result set. Therefore, the idea is to allow peers to start sending a message if and only if their local result sets coverage reaches a predefined threshold. With this result set coverage threshold, peers send intermediate results based on the improvement impact threshold obtained from the dynamic threshold function H define above.

5.5   Dynamic Threshold Algorithms

Our dynamic threshold approaches algorithms are based on the same principles as the static threshold ones. A generic algorithm for our dynamic threshold-based approaches is given in **Algorithm 3**. When a peer receives a new result set N from its children (or generates its own result set after local processing of a query), it first updates the set T_{cur} of its current top-k intermediate results with results in N (see line 2, Algorithm 3). If its current result set coverage cov is greater than the defined threshold result set coverage cov', then the peer computes the improvement threshold $delta$ using the dynamic function H and subsequently the improvement impact imp (see lines 3-5, Algorithm 3). If imp is greater than or equal to $delta$ or if there are no more children' results to wait for, then the peer sends the set $T_{tosend} = T_{cur} \setminus T_{old}$ to its parent and subsequently sets T_{curr} to T_{old} (see lines 6-9, Algorithm 3). Recall that T_{cur} is the set of the current top-k intermediate results and T_{old} is the top-k intermediate results sent so far to its parent.

6   Dealing with Peers Failures in ASAP

One main characteristics of P2P systems is the dynamic behaviour of peers. It may happen that some peers leave the system (or fail) during the query processing. As a result peers may become inaccessible in the result bubbling up phase. In this section, we deal with this problem.

6.1   Absence of Parent

In the query results bubbling up phase, each peer p bubbles up the results of peers in its subtree to its parent. It may happen that p's parent is inaccessible because it has left the system or failed. The question is which path to choose to bubble up p's intermediate results to the query originator. To deal with this problem a naive solution is that p sends its intermediate results directly to the query originator when p's parent is inaccessible. Recall that at the query forwarding phase the IP address and port of the query originator is communicated to all peers which have received the query. However the naive approach has some drawbacks. First, it may incur expensive merge of intermediates results at query originator which may be resource consuming. Second, by returning intermediate results of the peer whose parent is failed directly to the query originator, we reduce the capacity of peers to prune uninteresting intermediate results and this may increase significantly the volume of transferred data over the network.

Our solution to deal with the above mentioned problem is as following. Each peer p maintains locally for each active query q a list $QPath$ involving the addresses of peers (IP addresses and ports) in the path from the query originator to p in the q's forwarding tree. $QPath$ list is sorted by increasing positions of peers from the peer p in query q forwarding tree (the first item in this list is the parent of p, the second item is the grand parent of p, etc.). For constructing this list, each peer, including the query originator, adds its address to the query message before forwarding it to the neighbours. Thus, when a query message reaches a peer p, it contains the address of all parents of p.

In the phase of results bubble up when a peer detects that his parent (i.e first item in the list $QPath$) is inaccessible, the peer sends its new results to the next peer in the list $QPath$ which is reachable. Another problem which may happen is that a peer may leave the system without being able to send to its parent the results received so far from its children, and this may have serious impact in the accuracy of final top-k results. To overcome this problem we adopt the following approach. During the results bubbling up phase, when a peer finds that its parent is unreachable, it sends its current top-k results to the next available peer in the list $QPath$. Although, this technique can increase the volume of data transferred in highly dynamic environment it may improve significantly the accuracy of top-k results.

6.2   Adjustment of Peer' Local Result Set Coverage

When computing the local result set coverage, we must take into account the fact that a peer may change parent when its direct parent becomes inaccessible. Indeed, not taking this into account will result in overestimation of peers' result sets coverage which may affect the value of the impact of intermediate results and thereby reducing the ability of peers to bubble up good quality results as soon as possible. In this section, we present our technique for adjusting the local result set coverage which is based on updating sets $\mathcal{C}_1$ and $\mathcal{C}_2$ which each p_i maintains for each top-k query q and for each child p_j. Recall that $\mathcal{C}_1$ is a set of pairs (p_j, a) where $a \in \mathbb{N}$ is the number of peers in the sub-tree of peer p_j and $\mathcal{C}_2$ a set of pairs (p_j, e) where $e \in \mathbb{N}$ is the total number of peers in the sub-tree of peer p_j which have already processed locally the query.

To help peers to have a good estimation of their result set coverage when some peers become inaccessible, we modify as follows our approach for peers failures management presented previously. Each peer p_i maintains for its parent p_j and for each active query the latest values of the estimation of number of peers which have already processed the query in the sub-tree and the number of peers in its sub-tree it has sent to p_j. In the results bubbling up phase, when p_j is inaccessible, p_i inserts into its answer message to its new parent p_k (the first accessible peer in $QPath$ list) the following information: 1) the new and the latest (sent to p_j) values of the number of peers and the number of peers which have already processed locally in the sub-tree; 2) the address of the peer which is before p_k in the list $QPath$ (this peer is one of child of p_k in the query forwarding tree).

When a peer p_k receives an answer message of a query q from a peer p_j whose parent is inaccessible, it updates its estimation about the number of peers and the number of peers which have already processed locally q in the sub-tree of its direct child p_r which is declared as "inaccessible" by p_j (see lines 2-17, Algorithm 4). Then p_k activates a trigger to inform p_r (when it becomes accessible) that p_j is no longer in its sub-tree (see line 18, Algorithm 4).

7   Feedback Measures for Intermediate Results

Although it is important to provide good quality results as soon as possible to users, it is also interesting to associate "probabilistic guarantee" to the intermediate results allowing the user to know how far these results are from the final results. For example, we may wish to be able to give probabilistic guarantees, such as: "with probability γ, the current top-k results are likely to be the final top-k results". Our goal is to provide a mechanism to continuously compute these guarantees as results are bubbled up to the query originator. To do so, we compute two feedback measures which are returned to the user continuously: 1) the proportion of peers whose local results are already considered in the computation of the current top-k (we call this the proportion of contributor peers); 2) the probability of having the best k results in the current top-k results (we call

---

Algorithm 4. $result\_Coverage\_Adjustment(msg)$

```
     input : msg, an answer message of a query; C₁; C₂.
 1  begin
 2  |   if (change_Parent(msg)) then
 3  |   |   qid = msg.getQueryID();
 4  |   |   sender = msg.getAnswerSender();
 5  |   |   aₗ = getNbPeersSent(msg);
 6  |   |   eₗ = getNbAnsPeersSent(msg);
 7  |   |   p = getLastPeerInacessible(msg);
 8  |   |   if (eₗ > C₂.get(p)) then
 9  |   |   |   x₁ = C₂.get(p);
10  |   |   |   y₁ = C₁.get(p);
11  |   |   else
12  |   |   |   x₂ = C₂.get(p) − eₗ;
13  |   |   |   y₂ = C₁.get(p) − aₗ;
14  |   |   end
15  |   |   C₂.update(p, (x₁+x₂)/2);
16  |   |   y₂ = C₁.update(p, (y₁+y₂)/2);
17  |   end
18  |   propagateUpdate(queryId, Sender, p, aₗ, eₗ);
19  end
```

---

this the stabilization probability). In this section, we present how these feedback measures can be computed.

Note that our goal is not to provide approximative top-k result with probabilistic guarantees where at the end of the query execution approximative top-k result set is returned to user with a probability showing how this result is far from the exact top-k result. In our work, probabilistic guarantees are computed continuously during the execution of the query on intermediate results as in [5]. Notice that the work presented in [5] considered centralized databases where it is easier to get some information on data stored (e.g data distribution or scores distribution, number of data stored, etc.) and to collect statistics during query processing, in order to provide probabilistic guarantees for top-k intermediate results. Therefore, the approach proposed in [5] cannot easily be applied for unstructured P2P system where data are completely distributed (i.e. there is no centralized catalog).

7.1   Stabilization Probability

To be able to calculate the stabilization probability (i.e the probability that the current top-k is the exact top-k for a query q), we use the following information:

- the total number L of queried peers for the query q
- the total number M of data items shared by the L peers
- the number l of peers whose data items are taken into account in calculating the current top-k result of q
- the total number m of data items shared by the l peers

In Section 4.1, we presented how to estimate the parameter L and how to calculate l. Note that the calculation of m can be done by including in each answer message which a peer sends to its parent the number of data items which have already taken into account in the calculation of this response. However to be able to estimate M it is necessary to know the number l' of peers that are already known by the query initiator and the number m' of data items of those peers. The mechanism to calculate l' and m' works as follows. Each peer p_i maintains for each child p_j a set C_3 of pairs (p_j, c, d) where c is the number of peers which p_i knows in the sub-tree whose root is p_j and d the number of data items shared by the c peers. At the beginning of query processing, each peer sets $c = 0$ and $d = 0$. In the phase of result bubbled up, when a child p_j in $\psi(p_i, q)$ wants to send results to p_i, it inserts in the answer message the couple of values ($\sum\limits_{(p_j,c,d)\in C_3} c$, $\sum\limits_{(p_j,c,d)\in C_3} d$). Once p_i receives this message, it unpacks the message, gets these values and updates the set C_3.

The total number M of data items of all queried peers is then estimated using Formula 6.

$$M = \|D(p_0)\| + m' + \frac{\|D(p_0)\| + m'}{l' + 1} * (L - l' - 1) \tag{6}$$

where $D(p_0)$ is the number of data items shared by the query originator.

By assuming that the data distribution over peers is uniform, the probability P_k^m of finding the k best data items in the current top-k result is:

$$P_k^m = \frac{C_{M-k}^{m-k}}{C_M^m} \tag{7}$$

If l peers over L have already bubbled up their local results to query originator, the probability of having m data items on these l peers is:

$$P_l^m = \begin{vmatrix} 1 & if\ l = L \\ C_M^m \times (\frac{l}{L})^m \times (\frac{L-l}{L})^{M-m} & otherwise \end{vmatrix} \tag{8}$$

Knowing that l peers have already bubbled up their local results to query originator, the probability of having at least k data items is given by:

$$P_l^{\geq k} = \sum_{m=k}^{M} P_l^m \tag{9}$$

To find all the k best results in those l peers there must be at least k data items on these l peers and all the best results must be owned by these l peers. Thus the probability of having all the the top-k results in the current top-k result set is equal to:

$$P_l^{ktop} = \sum_{m=k}^{M} P_l^m \times P_k^m \tag{10}$$

To ensure a better estimation of the probability that the current top-k is the exact top-k in the case of peers failures, we have adopted the technique presented in Section 6 to readjust the estimation of all parameters used for calculating that probability.

7.2   Proportion of Contributor Peers

The proportion of contributor peers of a given current top-k results is the number of queried peers whose local results are already considered in the computation of that current top-k over the total number of queried peers. This proportion is equal to estimation of the query originator local result set coverage presented in Section 5. Thus, continuously we return the latter coverage to the user as the proportion of contributor peers.

7.3   Discussion

In some cases, it may happen that an unstructured P2P system is configured so that an issued query reaches all peers in the system (e.g by using very high ttl). In this case an efficient way to estimate the number of queried peers (i.e the network size) is to use gossip-based aggregation approach [19]. This approach relies on the following statement: if exactly one node of the system holds a value equal to 1, and all the other values are equal to 0, the average is $1/N$. The system size could thus be directly computed. To run this algorithm, an initiator should take the value equal to 1, and start gossiping; the reached nodes participate to the process by setting their value to 1. At each predefined cycle, each node in the network chooses one of its neighbors at random and swaps its estimation parameter (the network size and the number of shared data items). The contacted node does the same (push/pull heuristic of [19]). Both nodes then recompute their estimation as follows:

$$Estimation = \frac{Estimation + Neighbor's\_Estimation}{2}$$

By relying on gossip-based aggregation approach, we can also estimate the total number of data items shared by all peers in the system.

Notice that to provide correct estimations, this algorithm needs to wait a certain number of rounds to elapse before computing the size estimation; this period is the required time for the gossip to propagates in the whole overlay and for the values to converge. Notice that this method converge to the exact value in the stable system as demonstrated in [19]. Gossip protocols have been shown to provide exponentially fast convergence with low message transmission overhead as presented in [21].

8   Performance Evaluation

In this section, we evaluate the performance of ASAP through simulation using the PeerSim simulator [20]. This section is organized as follows. First, we describe our simulation setup, the metrics used for performance evaluation. Then, we study the effect of the number of peers and the number of results on the performance of ASAP, and show how it scales up. Next, we study the effect of the number of replicas on the performance of ASAP. We also study the effectiveness of our solution for providing probabilistic guarantees on the top-k results. After that, we study the effect of data distribution on the performance of ASAP. Finally, we investigate the effect of peers failures on the correctness of ASAP.

8.1 Simulation Setup

We implemented our simulation using the PeerSim simulator. PeerSim is an open source, Java based, P2P simulation framework aimed to develop and test any kind of P2P algorithm in a dynamic environment. It consists of configurable components and it has two types of engines: cycle-based and event-driven engine. PeerSim provides different modules that manage the overlay building process and the transport characteristics.

We conducted our experiments on a machine with a 2.4 GHz Intel Pentium 4 processor and 2GB memory. The simulation parameters are shown in Table 1. We use parameter values which are typical of P2P systems [15]. The latency between any two peers is a normally distributed random number with mean of 200 ms. Since users are usually interested in a small number of top results, we set $k = 20$ as default value. In our experiments we vary the network size from 1000 to 10000 peers. In order to simulate high heterogeneity, we set peers' capacities in our experiments, in accordance to the results in [15]. This work measures the peers capacities in the Gnutella system. Based on these results, we generate around 10% of low-capable, 60% of medium-capable, and 30% of high-capable peers. The high-capable peers are 3 times more capable than medium-capable peers and still 7 times more capable than low-capable ones.

In the context of our simulations each peer in the P2P system has a table $\mathcal{R}(data)$ in which attribute $data$ is a real value. The number of rows of $\mathcal{R}$ at each peer is a random number uniformly distributed over all peers greater than 1000 and less than 20000. Unless otherwise specified, we assume only one copy of each data item in our system (i.e. no data replication). We also ensure that there are not two different data items with the same score. In all our tests, we use the following simple query, denoted by q_{load} as workload:
SELECT val FROM $\mathcal{R}$ ORDER BY $F(\mathcal{R}.data, val)$ STOP AFTER k
The score $F(\mathcal{R}.data, val)$ is computed as:

$$\frac{1}{1 + |\mathcal{R}.data - val|}$$

In our simulation, we compare ASAP with Fully Distributed (FD) [2], a baseline approach for top-k query processing in unstructured P2P systems which works as follows. Each peer that receives the query, executes it locally (i.e. selects the k top scores), and waits for its children's results. After receiving all its children score-lists, the peer merges its k local top data items with those received from its children and selects the k top scores and sends the result to its parent.

Table 1. Simulation parameters

Parameters	Values
Latency	Normally distributed random number, $Mean = 200$ ms, $Variance = 100$
Number of peers	10,000 peers
Average degree of peers	4
ttl	9
k	20
Number of replicas	1

In our experiments, to evaluate the performance of ASAP comparing to FD, we use the following metrics:

(i) **Cumulative quality gap:** As defined in Section 3, is the sum of the quality difference between intermediate top-k result sets received until the stabilization time and the final top-k result set.

(ii) **Stabilization time:** We report on the stabilization time, the time of receiving all the final top-k results.

(iii) **Response time:** We report on the response time, the time the query initiator has to wait until the top-k query execution is finished.

(iv) **Communication cost:** We measure the communication cost in terms of number of answer messages and volume of data which must be transferred over the network in order to execute a top-k query.

(v) **Accuracy of results:** We define the accuracy of results as follows. Given a top-k query q, let V be the set of the k top results owned by the peers that received q, let V' be the set of top-k results which are returned to the user as the response of the query q. We denote the accuracy of results by ac_q and we define it as

$$ac_q = \frac{\|V \cap V'\|}{\|V\|}$$

(iv) **Total number of results:** We measure the total number of results as the number of results received by the query originator during query execution.

In our experimentation, we perform 30 tests for each experiment by issuing q_{load} 20 different times and we report the average of their results. Due to space limitations, we only present the main results of ASAP's dynamic threshold-based approaches denoted by ASAP-Dscore and ASAP-Drank. ASAP-Dscore uses a score-based improvement impact and ASAP-Drank a rank-based improvement impact. ASAP's dynamic threshold-based approaches have proved to be better than ASAP's static threshold-based approaches without being expensive in communication cost. In our all experiments, for ASAP-Dscore approach we use $H(x) = -0.2x + 0.2$ as dynamic threshold function and 0 as peer's local result set coverage threshold. In the case of Asap-Drank, we use $H(x) = -0.5x + 0.5$ as dynamic threshold function and 0.05 as peer's local result set coverage threshold.

8.2   Performance Results

Effect of Number of Peers. We study the effect of the number of peers on the performance of ASAP. For this, we ran experiments to study how cumulative quality gap, stabilization time, number of answer messages, volume of transferred data, number of intermediate results and response time increase with the addition of peers. Note that the other simulation parameters are set as in Table 1.

Figure 2(a) and 2(b) show respectively how cumulative quality gap and stabilization time increase with the number of peers. The results show that the cumulative quality gap of ASAP-Dscore and ASAP-Drank is always much smaller than that of FD, which means that ASAP returns quickly high quality results. The results also show that the stabilization time of ASAP-Dscore is always much smaller that of ASAP-Drank and that of FD. The reason is that ASAP-Dscore is score sensitive, so the final top-k results are obtained quickly.

Figure 2(c) shows that the total number of results received by the user increases with the number of peers in the case of ASAP-Dscore and ASAP-Drank while it is still constant in the case of FD. This is due to the fact that FD does not provide intermediate results to users. The results also show that the number of results received by the user in case of ASAP-Dscore is smaller than that of ASAP-Drank. The main reason is that ASAP-Dscore is score sensitive in contrast to ASAP-Drank.

Figure 2(d) and Figure 2(e) show that the number of answer messages and volume of transferred data increase with the number of peers. The results show that the number of answer messages and volume of transferred data of ASAP-Drank are always higher than those of ASAP-Dscore and FD. The results also show that the differences between ASAP-Dscore and FD's number of answer messages and volume of transferred data are not significant. The main reason is that ASAP-Dscore is score sensitive in contrast to ASAP-Drank. Thus, only high quality results are bubbled up quickly.

Figure 2(f) shows how response time increases with increasing the numbers of peers. The results show that the difference between ASAP-Dscore and FD response time is not significant. The results also show that the difference between ASAP-Drank and FD's response time increases slightly in favour of ASAP-Drank as the number of peers increases. The reason is that ASAP-Drank induces more network traffic than ASAP-Dscore and FD.

Effect of k. We study the effect of k, i.e. the number of results requested by the user, on the performance of ASAP. Using our simulator, we studied how cumulative quality gap, stabilization time and volume of transferred data evolve while increasing k from 20 to 100, with the other simulation parameters set as in Table 1. The results (see Figure 3(a), Figure 3(b)) show that k has very slight impact on cumulative quality gap and stabilization time of ASAP-Dscore and ASAP-Drank. The results (see Figure 3(c)) also show that by increasing k, the volume of transferred data of ASAP-Dscore and ASAP-Drank increase less than

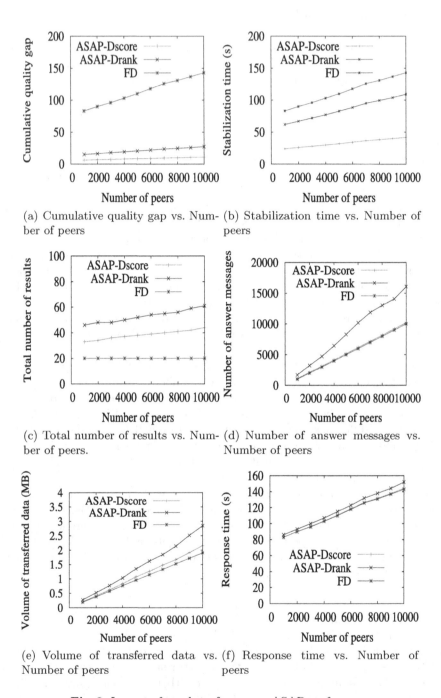

Fig. 2. Impact of number of peers on ASAP performance

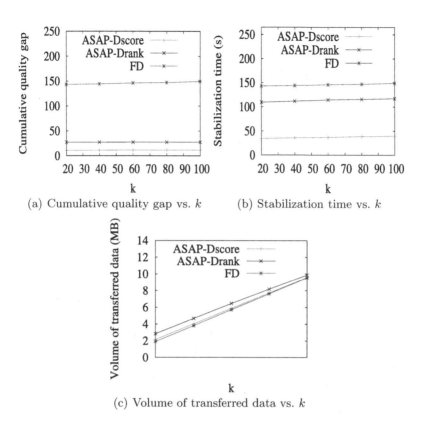

(a) Cumulative quality gap vs. k

(b) Stabilization time vs. k

(c) Volume of transferred data vs. k

Fig. 3. Impact of k on ASAP performance

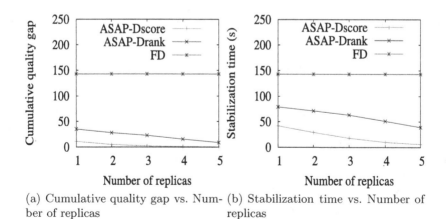

(a) Cumulative quality gap vs. Number of replicas

(b) Stabilization time vs. Number of replicas

Fig. 4. Impact of data replication on ASAP performance

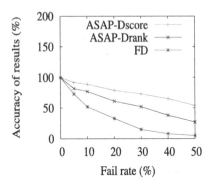

Fig. 5. Accuracy of results vs. fail rate

that of FD. This is due to the fact that ASAP-Dscore and ASAP-Drank prune more intermediate results when k increases.

Data Replication. Replication is widely used in unstructured P2P systems to improve search or achieve availability. For example, modern unstructured overlays like BubbleStorm [29] use large number of replicas for each object placed in the overlay to improve their search algorithms.

We study the effect of the number of replicas, which we replicate for each data (uniform replication strategy [14]), on the performance of ASAP. Using our simulator, we studied how cumulative quality gap and stabilization time evolve while increasing the number of replicas, with the other simulation parameters set as in Table 1. The results (see Figure 4(a) and Figure 4(b)) show that increasing the number of replicas for ASAP and FD decrease ASAP-Dscore and ASAP-Drank's cumulative quality gap and stabilization time. However, FD's cumulative quality gap and stabilization time are still constant. The reason is that ASAP returns quickly the results having high quality in contrast to FD which returns results only at the end of query execution. Thus, if we increase the number of replicas, ASAP finds quickly the results having high scores.

Effectiveness of Our Solution for Computing "Probabilistic Guarantees". In this section, we study the effectiveness of the proposed solution in Section 7 for computing the probabilistic guarantees, by comparing them with optimal values. For this, we ran experiments to study how our probabilistic guarantees values evolve comparing to the optimal (i.e real) values during the query execution in the case of ASAP-Dscore. Figure 6(a) and Figure 6(b) show that the difference between our probabilistic guarantees values and the exact values is very slight. The results also show that our probabilistic guarantees values converge to exact values and this before the end of the execution of the query execution. This means that our solution provides reliable guarantees for the user on the intermediate results.

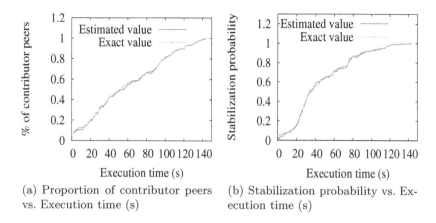

(a) Proportion of contributor peers vs. Execution time (s)

(b) Stabilization probability vs. Execution time (s)

Fig. 6. Effectiveness of our solution for computing probabilistic guarantees

Data Distribution. In this section, we study the effect of data distribution on the performance of our top-k query processing solution. Often relevant data items are grouped together, stored on a group of neighbouring peers. If these groups of peers have some good data objects for top-k, they become the sweet region in the network that can contribute a lot to a final top-k. To study the effect of data distribution on the performance of ASAP, we randomly distribute the top-k data items of our test bed queries respectively on 4, 6, 8 and 10 peers of P2P system and the other data (i.e which are not in the top-k results) uniformly over all the peers of the system. Using our simulator, we studied how cumulative quality gap, stabilization time and volume of transferred data evolve while only 4, 6, 8 and 10 peers of the P2P system store the top-k results with the other simulation parameters set as in Table 1. The results (see Figure 7(a) and Figure 7(b)) show that ASAP can take advantage of grouped data distribution to provide quickly high quality results to users in contrast to FD. The results (see Figure 7(c)) also show that in the case of ASAP, the higher the top-k data items are grouped together, the smaller is the volume of transferred data over the network, while this volume is constant in the case of FD.

Effect of Peers Failures. In this section, we investigate the effect of peers' failures on the accuracy of top-k results of ASAP. In our tests, we vary the value of fail rate and investigate its effect on the accuracy of top-k results. Figure 5 shows accuracy of top-k results for ASAP-Dscore, ASAP-Drank and FD while increasing the fail rate, with the other parameters set as in Table 1. Peers' failures have less impact on ASAP-Dscore and ASAP-Drank than FD. The reason is that ASAP-Dscore and ASAP-Drank return the high-score results to the user as soon as possible. However, when increasing the fail rate in FD, the accuracy of top-k results decreases significantly because some score-lists are lost. Indeed, in FD, each peer waits for results of its children so in the case of a peer failure, all the score-lists received so far by that peer are lost.

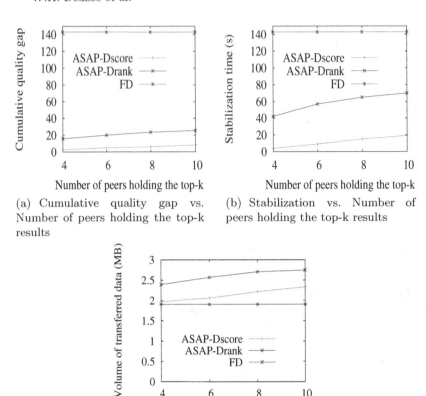

(a) Cumulative quality gap vs. Number of peers holding the top-k results

(b) Stabilization vs. Number of peers holding the top-k results

(c) Volume of transferred data vs. Number of peers holding the top-k results

Fig. 7. Impact of data distribution on ASAP performance

9   Related Work

Efficient processing of top-k queries is both an important and hard problem that is still receiving much attention. Several papers have dealt with top-k query processing in centralized database management systems [9,18,27,24]. In distributed systems [10,16,7,31,33], previous work on top-k processing has focused on vertically distributed data over multiple sources, where each source provides a ranking over some attributes. Some of the proposed approaches, such as recently [3], try to improve some limitations of the Threshold Algorithm (TA) [13]. Following the same concept, there exist some previous work for top-k queries in P2P over vertically distributed data. In [8], the authors propose an algorithm called "Three-Phase Uniform Threshold" (TPUT) which aims at reducing communication cost by pruning away intelligible data items and restricting the number of round-trip messages between the query originator and other nodes. Later, TPUT was improved by KLEE [22] that uses the concept of bloom filters to reduce the data communicated over the network upon processing top-k queries. It brings significant performance benefits with small penalties in result precision. However, theses approaches assume that data is vertically distributed over the nodes whereas we deal with horizontal data distribution.

For horizontally distributed data, there has been little work on P2P top-k processing. In [2], the authors present FD, a fully distributed approach for top-k query processing in unstructured P2P systems. We have briefly introduced FD in section 8.1.

PlanetP [11] is the content addressable publish/subscribe service for unstructured P2P communities up to ten thousand peers. PlanetP uses a gossip protocol to replicate global compact summaries of content (term-to-peer mappings) which are shared by each peer. The top-k processing algorithm works as follows. Given a query q, the query originator computes a relevance ranking (using the global compact summary) of peers with respect to q, contacts them one by one from top to bottom of ranking and asks them to return a set of their top-scored document names together with their scores. However, in a large P2P system, keeping up-to-date the replicated index is a major problem that hurts scalability.

In [6], the authors present an index routing based top-k processing technique for super-peer networks organized in an HyperCuP topology which tries to minimize the number of transfer data. The authors use statistics on queries to maintain the indexes built on super-peers. However, the performance of this technique depends on the query distribution.

In [32], the authors present SPEERTO, a framework that supports top-k query processing in super-peer networks by using a skyline operator. In SPEERTO, for a maximum of K, denoting an upper bound on the number of results requested by any top-k query ($k \leq K$), each peer computes its K-skyband as a pre-processing step. Each super peer maintains and aggregates the K-skyband sets of its peers to answer any incoming top-k query. The main drawback of this approach is that each join or leave of peer may induce the recomputing of all super-peers K-skyband. Although these techniques are very good for super-peers systems, they cannot apply efficiently for unstructured P2P systems, since there may be no peer with high reliability and computing power.

Zhao *et al.* [34] use a result caching technique to prune network paths and answer queries without contacting all peers. The performance of this technique depends on the query distribution. They assume acyclic networks, which is restrictive for unstructured P2P systems.

10   Conclusion

In this paper we deal with as-soon-as-possible top-k query processing in P2P systems. We proposed a formal definition for as-soon-as-possible top-k query processing by introducing two novels notions: stabilization time and cumulative quality gap. We presented ASAP, a family of algorithms which uses a threshold-based scheme that considers the score and the rank of intermediate results to return quickly the high quality results to users. We validated ASAP through implementation and extensive experimentation. The results show that ASAP significantly outperforms baseline algorithms by returning final top-k result to users in much better times. Finally, the results demonstrate that in the presence of peers' failures, ASAP provides approximative top-k results with good accuracy, unlike baseline algorithms.

References

1. Akbarinia, R., Martins, V., Pacitti, E., Valduriez, P.: Design and Implementation of Atlas P2P Architecture. In: Global Data Management, 1st edn. IOS Press (2006)
2. Akbarinia, R., Pacitti, E., Valduriez, P.: Reducing network traffic in unstructured p2p systems using top-k queries. Distributed and Parallel Databases 19(2-3), 67–86 (2006)
3. Akbarinia, R., Pacitti, E., Valduriez, P.: Best position algorithms for top-k queries. In: Proceedings of Int. Conf. on Very Large Data Bases (VLDB), pp. 495–506 (2007)
4. Androutsellis-Theotokis, S., Spinellis, D.: A survey of peer-to-peer content distribution technologies. ACM Computing Surveys 36(4), 335–371 (2004)
5. Arai, B., Das, G., Gunopulos, D., Koudas, N.: Anytime measures for top-k algorithms. In: Proceedings of Int. Conf. on Very Large Data Bases (VLDB), pp. 914–925 (2007)
6. Balke, W.-T., Nejdl, W., Siberski, W., Thaden, U.: Progressive distributed top k retrieval in peer-to-peer networks. In: Proceedings of Int. Conf. on Data Engineering (ICDE), pp. 174–185 (2005)
7. Bruno, N., Gravano, L., Marian, A.: Evaluating top-k queries over web-accessible databases. In: Proceedings of Int. Conf. on Data Engineering (ICDE), pp. 369–380 (2002)
8. Cao, P., Wan, Z.: Efficient top-k query calculation in distributed networks. In: Proceedings of Annual ACM Symposium on Principles of Distributed Computing (PODC), pp. 206–215 (2004)
9. Chaudhuri, S., Gravano, L.: Evaluating top-k selection queries. In: Proceedings of Int. Conf. on Very Large Databases (VLDB), pp. 397–410 (1999)
10. Chaudhuri, S., Gravano, L., Marian, A.: Optimizing top-k selection queries over multimedia repositories. IEEE Transactions on Knowledge Data Engineering 16(8), 992–1009 (2004)
11. Cuenca-Acuna, F.M., Peery, C., Martin, R.P., Nguyen, T.D.: Planetp: Using gossiping to build content addressable peer-to-peer information sharing communities. In: Proceedings of IEEE Int. Symp. on High-Performance Distributed Computing (HPDC), pp. 236–249 (2003)
12. Dedzoe, W.K., Lamarre, P., Akbarinia, R., Valduriez, P.: Asap top-k query processing in unstructured p2p systems. In: Proceedings of IEEE Int. Conf on Peer-to-Peer Computing (P2P), pp. 187–196 (2010)
13. Fagin, R., Lotem, A., Naor, M.: Optimal aggregation algorithms for middleware. In: Proceedings of Symposium on Principles of Database Systems (PODS), pp. 102–113 (2001)
14. Feng, G., Jiang, Y., Chen, G., Gu, Q., Lu, S., Chen, D.: Replication strategy in unstructured peer-to-peer systems. In: Proceedings of IEEE International Parallel and Distributed Processing Symposium (IPDPS), pp. 1–8 (2007)
15. Gummadi, P.K., Saroiu, S., Gribble, S.D.: A measurement study of napster and gnutella as examples of peer-to-peer file sharing systems. Computer Communication Review 32(1), 82 (2002)

16. Güntzer, U., Balke, W.-T., Kießling, W.: Optimizing multi-feature queries for image databases. In: Proceedings of Int. Conf. on Very Large DataBases (VLDB), pp. 419–428 (2000)
17. Hose, K., Karnstedt, M., Sattler, K.-U., Zinn, D.: Processing top-n queries in p2p-based web integration systems with probabilistic guarantees. In: Proceedings of International Workshop on web and databases (WebDB), pp. 109–114 (2005)
18. Hristidis, V., Koudas, N., Papakonstantinou, Y.: Prefer: A system for the efficient execution of multi-parametric ranked queries. In: Proceedings of ACM. Int. Conf. on Management of Data (SIGMOD), pp. 259–270 (2001)
19. Jelasity, M., Montresor, A.: Epidemic-style proactive aggregation in large overlay networks. In: Int. Conference on Distributed Computing Systems (ICDCS), pp. 102–109 (2004)
20. Jelasity, M., Montresor, A., Jesi, G.P., Voulgaris, S.: The Peersim simulator, http://peersim.sf.net
21. Kempe, D., Dobra, A., Gehrke, J.: Gossip-based computation of aggregate information. In: Symposium on Foundations of Computer Science (FOCS), pp. 482–491 (2003)
22. Michel, S., Triantafillou, P., Weikum, G.: Klee: A framework for distributed top-k query algorithms. In: Proceedings of Int. Conf. on Very Large Data Bases (VLDB), pp. 637–648 (2005)
23. Ooi, B.C., Shu, Y., Tan, K.-L.: Relational data sharing in peer-based data management systems. SIGMOD Record 32(3), 59–64 (2003)
24. Qin, L., Yu, J.X., Chang, L.: Diversifying top-k results. PVLDB 5(11), 1124–1135 (2012)
25. Ramaswamy, L., Chen, J., Parate, P.: Coquos: Lightweight support for continuous queries in unstructured overlays. In: Proceedings of IEEE International Parallel and Distributed Processing Symposium (IPDPS), pp. 1–10 (2007)
26. Schmid, S., Wattenhofer, R.: Structuring unstructured peer-to-peer networks. In: Proceedings of IEEE Int. Conf. on High Performance Computing (HiPC), pp. 432–442 (2007)
27. Shmueli-Scheuer, M., Li, C., Mass, Y., Roitman, H., Schenkel, R., Weikum, G.: Best-effort top-k query processing under budgetary constraints. In: Proceedings of Int. Conf. on Data Engineering (ICDE), pp. 928–939 (2009)
28. Tatarinov, I., Ives, Z.G., Madhavan, J., Halevy, A.Y., Suciu, D., Dalvi, N.N., Dong, X., Kadiyska, Y., Miklau, G., Mork, P.: The piazza peer data management project. SIGMOD Record 32(3), 47–52 (2003)
29. Terpstra, W.W., Kangasharju, J., Leng, C., Buchmann, A.P.: Bubblestorm: resilient, probabilistic, and exhaustive peer-to-peer search. In: SIGCOMM, pp. 49–60 (2007)
30. Tsoumakos, D., Roussopoulos, N.: Analysis and comparison of p2p search methods. In: Proceedings of Int. Conf. on Scalable Information Systems (Infoscale), p. 25 (2006)
31. Vlachou, A., Doulkeridis, C., Nørvåg, K.: Distributed top-k query processing by exploiting skyline summaries. Distributed and Parallel Databases 30(3-4), 239–271 (2012)
32. Vlachou, A., Doulkeridis, C., Nørvåg, K., Vazirgiannis, M.: On efficient top-k query processing in highly distributed environments. In: Proceedings of ACM. Int Conf. on Management of Data (SIGMOD), pp. 753–764 (2008)
33. Ye, M., Lee, W.-C., Lee, D.L., Liu, X.: Distributed processing of probabilistic top-k queries in wireless sensor networks. IEEE Trans. Knowl. Data Eng. 25(1), 76–91 (2013)
34. Zhao, K., Tao, Y., Zhou, S.: Efficient top-k processing in large-scaled distributed environments. Data and Knowledge Engineering 63(2), 315–335 (2007)

Self-stabilizing Consensus Average Algorithm in Distributed Sensor Networks

Jacques M. Bahi[1], Mohammed Haddad[2],
Mourad Hakem[1], and Hamamache Kheddouci[2]

[1] DISC Laboratory, Femto-ST - UMR CNRS, Université de Franche-Comté, France
[2] LIRIS Laboratory, UMR CNRS 5205, Université de Lyon 1, F-69622, France
{Mourad.Hakem,Jacques.Bahi}@lifc.univ-fcomte.fr,
{Mohammed.Haddad,Hamamache.Kheddouci}@univ-lyon1.fr

Abstract. One important issue in sensor networks that has received renewed interest recently is average consensus, i.e., computing the average of n sensor measurements, where nodes iteratively exchange data with their neighbors and update their own data accordingly until reaching convergence to the right parameters estimate. In this paper, we introduce an efficient self-stabilizing algorithm to achieve/ensure the convergence of node states to the average of the initial measurements of the network. We prove that the convergence of the fusion process is finite and express an upper bound of the actual number of moves/iterations required by the algorithm. This means that our algorithm is guaranteed to reach a stable situation where no load will be sent from one sensor node to another. We also prove that the load difference between any two sensor nodes in the network is within $\frac{\varepsilon}{D} \times \lfloor \frac{D+1}{2} \rfloor < \varepsilon$, where ε is the prescribed global equilibrium threshold (this threshold is given by the system) and D is the diameter of the network.

1 Introduction

Recent years have witnessed significant advances in wireless sensor networks which emerge as one of the most promising technologies for the 21st century [1]. In fact, they present huge potential in several domains ranging from health care applications to military applications. Distributed in irregular patterns across remote and often hostile environments, sensor nodes will autonomously aggregate into collaborative and asynchronous communication mode. Indeed, the asynchronous mode presents the major advantages of allowing more flexible communication schemes. They are less sensitive to the communication delays and to their variations. Moreover, they also present some tolerance to the loss of data messages since that losses do not prevent the progression of the fusion process on both the sender and destination nodes.

In general, the primary objective of a wireless sensor network is to collect data from the monitored area and to transmit it to a base station (sink) for processing. During this phase, resource failures are more likely to occur and can have an adverse effect on the application. Hence, they must be robust and survivable despite individual node and link failures [2, 3, 4, 5, 6]. The advent of wireless sensor networks and its conception constraints, have posed a number of research challenges to the networking and

A. Hameurlain et al. (Eds.): TLDKS IX, LNCS 7980, pp. 28–41, 2013.

distributed computation communities. A problem that has received renewed interest recently is **average consensus**. It computes iteratively the global average of distributed measures in a sensor network by using only local communications. Distributed average consensus, in ad hoc networks, is an important issue in distributed agreement and synchronization problems [7] and is also a central topic for load balancing (with divisible tasks) in parallel computing [8, 9]. More recently, it has also found applications in distributed coordination of mobile autonomous agents and distributed data fusion in sensor networks [10, 11, 12, 13].

In the literature, this problem has been formulated and studied in various ways. The first approaches were based on flooding. For instance, in [14], each sensor node broadcasts all its stored and received data to its neighbors. After some times, each node will hold all the data of the network and acts as a fusion center to compute the estimate of the unknown parameter. In [15, 16, 17], the authors compute the average of the sensor measurements combined with local Kalman filtering and/or mobile agents. The works developed in [14, 18, 19] consist of distributed linear iterations, where each sensor updates its current state by a weighted fusion of its current neighbors' states (which are distorted when they reach it) and these fusion weights decrease to zero in an appropriate way, as time progresses. Other authors consider some practical issues in sensor networks such as fault tolerance and asynchronism. For instance, some works compute the average while taking into account link failures [20], other works study the consensus problem into asynchronous environment [21, 22] while considering communication delays, or from the energy point of view by minimizing the number of iterations [19].

To the best of our knowledge, none of the above approaches is able to give an analytical bound of the actual number of moves/iterations required by the algorithm, nor to improve the upper bound for the load difference (upper bounded by the diameter of the topology) between any two sensor nodes in the final load balanced distribution. In this paper, we present an efficient self-stabilizing algorithm to tackle the problem of distributed data fusion in large-scale sensor networks. This study differs from previous works for the following reasons:

- We express an upper bound of the actual number of moves/iterations required by the algorithm to ensure the convergence of node states to the average of the initial measurements of the network. More precisely, we prove that there exists an upper bound of the convergence time beyond which all the sensor nodes in the network neither receive nor send any amount of load and, therefore, achieve a stable balanced state.
- We improve the load difference between any two sensor nodes in the network which is within $\frac{\varepsilon}{D} \times \lfloor \frac{D+1}{2} \rfloor$ rather than ε, where ε is the prescribed global equilibrium threshold (this threshold is given by the system) and D is the diameter of the network.
- Unlike earlier methods, we use a new concept of Self-Stabilization to achieve the convergence of the system to a final balanced load state.

In a self-stabilizing model [23, 24, 25, 26], each vertex has only a partial view of the system, called the *local state*. The vertex's local state include the state of the vertex itself and the state of its neighborhood. The union of the local states of all the vertices gives the *global state* of the system. Based on its local state, a vertex can decide to make *a move*. Then, self-stabilizing algorithms are given as a set of rules of the form [**If** $p(i)$ **Then** M], where $p(i)$ is a predicate and M is a move. $p(i)$ is true when state of the vertex i is locally illegitimate. In this case, the vertex i is called a *privileged/active* vertex. A vertex executes the algorithm as long as it is active (at least one predicate is true).

The rest of the paper is organized as follows. After some definitions and notations in Section 2.1, we present in Sections 2 the design and analysis of the proposed self-stabilizing algorithm and give the corresponding proofs. To evaluate the behavior of the proposed algorithm, we provide in Section 3 some results through simulations that we conducted on NS2 (Network Simulator 2). Finally we give some concluding remarks in Section 4 and 5.

2 Self-stabilizing Consensus Average Algorithm

In this section, we give a self-stabilizing algorithm for computing the consensus average in a wireless sensor network under a serial, or central, scheduler. Nevertheless, there exist algorithms that make any self-stabilizing algorithm using the central scheduler operate under the distributed one [27, 28, 29, 30, 31]. We also assume a composite read/write atomicity. We begin by giving fundamentals and a description of our algorithm then we focus on the legitimate state formulation as well as the local information at the nodes. After that, we present the algorithm which consists in only one rule and give the proofs.

2.1 Fundamentals

A sensor network is modeled as a connected undirected graph $G = (V, E)$. The set of nodes is denoted by V (the set of vertices), and the links between nodes by E (the set of edges). The nodes are labeled $i = 1, 2, \ldots, n$, and a link between nodes i and j is denoted by (i, j). The set of neighbors of node i is denoted by $N_i = \{j \in V \mid (i, j) \in E\}$, and the degree (number of neighbors) of node i $\eta_i = |N_i|$. Each node takes initial measurement z_i, for the sake of simplicity, let us suppose that $z_i \in \mathbb{R}$. Then, z will refer to the vector whose ith component is z_i. Each node on the network also maintains a dynamic state $x_i(t) \in \mathbb{R}$ which is initially set to $x_i(0) = z_i$. Intuitively each node's state $x_i(t)$ is its current estimate of the average value $\sum_{i=1}^{n} z_i/n$. The goal of the averaging algorithm, is to let all the states $x_i(t)$ go to the average $\sum_{i=1}^{n} z_i/n$, as $t \to \infty$. Throughout the paper, we use the terms *scalar* and *load* interchangeably.

In our framework, instead of reaching $\sum_{i=1}^{n} z_i/n$, when $t \to \infty$, we ensure reaching $\sum_{i=1}^{n} z_i/n \pm \varepsilon$ but in a finite time. Where ε is the prescribed global equilibrium threshold.

2.2 Outline of the Algorithm

In order to reach, in a fully distributed way, the global consensus average, we draw inspiration from a natural phenomenon that fits well as a model for our problem. This phenomenon is the *communicating vessels*. In fact, one can see that by considering nodes the network as similar vessels all filled with some amount of water (the sensed value), then by making all the vessels communicating we will obtain, after stabilization, the same amount of water in all vessels. This amount is actually the global average (see Figure 1).

To model the behavior of the transfer of water from a vessel to another, we also act as in the natural phenomenon; that is the vessels with low amount of water create a depression and aspirate water from more loaded neighbors until the equilibrium is reached. Hence, there will be streams of water circulating between the vessels as a vessel could aspirate and be aspirated at the same time. In our model, we transfer an atomic quantity ϵ from a highly loaded node to a less loaded node until they reach the equilibrium. This transfer is supposed to be performed by some atomic transaction mechanism that could be called by our algorithm. Thus, the atomic transaction algorithm will be composed with our algorithm [32].

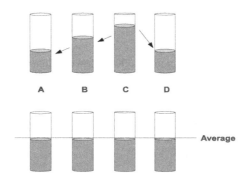

Fig. 1. Communicating vessels

2.3   Global Legitimate State

Let $G = (V, E)$ the graph modeling the sensor network. The algorithm should converge to a state where all node reach the same value representing the consensus average. However, we admit some error in the precision; that is two nodes should reach the same value according to some error ε. The legitimate state of the network is then expressed as follows:

$$\forall i, j \in V : |x_i - x_j| \leq \varepsilon \tag{1}$$

where ε is the prescribed global equilibrium threshold. This threshold is given by the system. We first prove that the Statement (1) ensures that every node in the network has reached the consensus average within a certain error e but always within the threshold ε.

Theorem 1. *Let $G = (V, E)$ be a graph such that $|V| = n$.*

$$(\forall i, j \in V : |x_i - x_j| \leq \varepsilon) \implies (\forall i \in V : x_i = \frac{\sum_j x_j}{n} \pm e_i \wedge e_i \leq \varepsilon)$$

Proof. Since all the vertices are holding the same value according to a given error ε, we have:

$$
\begin{aligned}
&(\forall i, j \in V : |x_i - x_j| \leq \varepsilon) \\
\Leftrightarrow &(\forall i, j \in V : -\varepsilon \leq x_i - x_j \leq \varepsilon) \\
\Leftrightarrow &(\forall i, j \in V : x_j - \varepsilon \leq x_i \leq x_j + \varepsilon) \\
\Leftrightarrow &(\forall i, j \in V : x_i = x_j \pm e_i \wedge e_i \leq \varepsilon) \\
\Rightarrow &(\forall i \in V : \sum_j x_i = \sum_j x_j \pm \sum_j e_i \wedge e_i \leq \varepsilon) \\
\Rightarrow &(\forall i \in V : n \times x_i = \sum_j x_j \pm n \times e_i \wedge e_i \leq \varepsilon) \\
\Rightarrow &(\forall i \in V : x_i = \frac{\sum_j x_j}{n} \pm e_i \wedge e_i \leq \varepsilon) \quad\quad\quad \square
\end{aligned}
$$

2.4 Local Information

Every node i in the network has to maintain the following data structure:

- x_i: the scalar value at node i.
- N_i: the set of neighbors of node i.
- σ: the local equilibrium threshold.

Fig. 2. The threshold σ

The threshold σ has to be chosen such that the transitive difference between nodes will never exceed the real threshold ε (see Figure 2). In fact, let's suppose three vertices a, b and c such that a is a neighbor of b which also a neighbor of c but a an c aren't neighbors. If the difference between the values x_a and x_b is less than σ and the difference between the values x_b and x_c is less than σ then what could we say about the difference between the values x_a and x_c ? Hence, the threshold σ is defined according to the diameter of the network D. Actually, by setting $\sigma \leq \varepsilon/D$, we obtain a sufficient condition on vertices to ensure the global threshold. The deployment knowledge of sensor networks is often used to get better performance. Indeed, in [33] deployment knowledge like the number of nodes and the diameter of the network is addressed.

2.5   The Algorithm

As mentioned above, the algorithm consists in only on rule that

---
2.1. The rule R_1 : Local equilibrium

R_1: Transfers σ from a neighbor j to i if j is more loaded than i.

If $\exists j \in N_i : x_j - x_i > \sigma$ **Then**
  $Transfer(x_j, x_i)$
End If

With

2.2. Transfer Transaction Procedure

$Transfer(x_j, x_i)$
$x_j = x_j - \sigma$
$x_i = x_i + \sigma$

2.6   Convergence Proof

Let $G = (V, E)$ the graph modeling the sensor network, with $|V| = n$ and $|E| = m$. In the following, we consider a discrete time where every move increments the time t by 1. Let $Max(t)$ be the maximum value in the network at the time t and respectively $Min(t)$ be the minimum value.

Lemma 1. $\forall t, Max(t) \geq Max(t + 1)$ *(respectively, $\forall t, Min(t) \leq Min(t + 1)$).*

Proof. the proof is straightforward since we transfer an atomic quantity σ from a highly loaded node to a less loaded node.   □

Lemma 2. *If the system is unstable, that is the Statement (1) is false, then we have $Max(t) < Max(t + \Delta t)$ such that Δt is within $O(n)$ moves.*

Proof. The worst case is when only one vertex is not in the equilibrium (consider it to be the black vertex in Figure 3). Since all other vertices are in equilibrium, all of them are holding the maximum value. Hence, in the worst case, the transfer stream will be formed by all the vertices in the network as a Hamiltonian path. This produces that the Max (rsp. Min) value will be decremented (rsp. incremented) by at least σ within $O(n)$ moves.   □

Theorem 2. *The algorithm described by the rule R_1 converges within $O\left(\dfrac{Max(0) - Min(0)}{\sigma} \times n\right)$ moves.*

Proof. By the previous lemmas, we have seen that Max value is decremented by at least σ within $O(n)$ moves (rsp. for Min). The worst case here is when the average is close to one of the extremal values either $Max(0)$ or $Min(0)$. Hence, $O\left(\dfrac{Max(0) - Min(0)}{\sigma}\right)$ transfers will be needed to reach the average. Since every transfer could cost $O(n)$ moves, we obtain that the algorithm converges within $O\left(\dfrac{Max(0) - Min(0)}{\sigma} \times n\right)$.   □

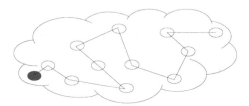

Fig. 3. Worst case of load transfer

2.7   Improvement of the Algorithm

We propose now to introduce a new rule to the algorithm. This rule aims to improve the global equilibrium of the network while proceeding only on local information. The rule is as follows:

2.3. Neighborhood Equilibrium

R_2: Transfers σ from a neighbor k to a neighbor l if k is more loaded than l.
If $(\forall j \in N_i : |x_j - x_i| \leq \sigma) \wedge (\exists k, l \in N_i : x_k - x_l > \sigma)$ **Then**
   $Transfer(x_k, x_l)$
End If

With

2.4. Transfer procedure

$Transfer(x_k, x_l)$
$x_k = x_k - \sigma$
$x_l = x_l + \sigma$

Observe however that this second transfer procedure will require distance two knowledge.

Theorem 3. *The load difference between any two sensor nodes in the network is within* $\frac{\varepsilon}{D} \times \lfloor \frac{D+1}{2} \rfloor < \varepsilon$, *where ε is the prescribed global equilibrium threshold (this threshold is given by the system) and D is the diameter of the network.*

Proof. To prove this bound, consider a linear chain graph of n nodes ($D = n - 1$) arranged with ascending order of their loads x_i, $1 \leq i \leq n$ along a line. If all nodes are in the following state:

$$x_1 = \frac{\varepsilon}{n-1} < x_2 = 2 \times \frac{\varepsilon}{n-1}$$
$$< \cdots <$$
$$x_{i-1} = (i-1) \times \frac{\varepsilon}{n-1} < x_i = i \times \frac{\varepsilon}{n-1}$$
$$< \cdots <$$
$$x_{n-1} = \varepsilon < x_n = n \times \frac{\varepsilon}{n-1}$$

Then, for this configuration, the only rule that can be executed is rule 2. For the sake of simplicity, assume that nodes, with index $i \mod 2 = 0$, will be activated for rule 2. Thus, we get

$$x_1 = x_2 = x_3 = 2 \times \frac{\varepsilon}{n-1}$$
$$< \cdots <$$
$$x_{i-1} = x_i = x_{i+1} = i \times \frac{\varepsilon}{n-1}$$
$$< \cdots <$$
$$x_{n-2} = x_{n-1} = x_n = \varepsilon$$

Similarly, for this configuration, the only rule that can be executed is rule 1 and for the sake of simplicity, assume that the involved nodes $(3, 6, \ldots n - 3)$ will be activated for rule 1. Thus, we get:

$$x_1 = x_2 = 2 \times \frac{\varepsilon}{n-1} < x_3 = x_4 = 3 \times \frac{\varepsilon}{n-1}$$
$$< \cdots <$$
$$x_{i-1} = x_i = i \times \frac{\varepsilon}{n-1} < x_{i+1} = x_{i+2} = (i+1) \times \frac{\varepsilon}{n-1}$$
$$< \cdots <$$
$$x_{n-3} = x_{n-2} = \varepsilon - \frac{\varepsilon}{n-1} < x_{n-1} = x_n = \varepsilon$$

Now, by alternating the execution of the two rules, there will be streams of load circulating between the nodes as a node could aspirate and be aspirated at the same time. Hence, this process is repeated until reaching the configuration case where nodes on the line are in the ascending order by an increment of $\frac{\varepsilon}{n-1}$ with at least two adjacent nodes which have the same load value. Formally:

$$\forall i, \forall k, l \in N_i, \ k \notin N_l : x_k - x_l \leq \sigma$$

In this case, all nodes will never again execute rule 2. This means, that all nodes reach their stable state where no load will be sent from one sensor node to another.

This configuration case can be viewed as a splitting of the initial linear chain graph into a new linear chain of virtual nodes, where each virtual node contains at least two nodes with the same load value. The virtual nodes along a new chain graph are in the ascending order by an increment of $\frac{\varepsilon}{n-1}$.

Thus, it follows that for $n \geq 2$ the load difference between any two sensor nodes in the linear chain graph is within

$$\frac{\varepsilon}{n-1} \times \left\lfloor \frac{n}{2} \right\rfloor = \frac{\varepsilon}{D} \times \left\lfloor \frac{D+1}{2} \right\rfloor < \varepsilon$$

□

Theorem 4. *The bound* $\frac{\varepsilon}{D} \times \left\lfloor \frac{D+1}{2} \right\rfloor$ *is attainable.*

Proof. To see that this bound is really attainable, consider a linear chain graph of $n = 6$, a non negative integer $\varepsilon = 5$ and

$$\sigma = \frac{\varepsilon}{D} = \frac{5}{5} = 1 = |x_{i+1} - x_i|, \ 1 \leq i \leq n - 1$$

By alternating the execution of the two rules, we obtain the final stable situation of loads

$$x_1 = 2 < x_2 = x_3 = 3 < x_4 = x_5 = 4 < x_6 = 5$$

with the difference of

$$\frac{\varepsilon}{n-1} \times \left\lfloor \frac{n}{2} \right\rfloor = \frac{\varepsilon}{D} \times \left\lfloor \frac{D+1}{2} \right\rfloor = \frac{5}{5} \times \left\lfloor \frac{5+1}{2} \right\rfloor = 3 < \varepsilon = 5$$

$\square$

Theorem 5. *For a non negative integer load balancing problem, the load difference between any two sensor nodes in the network is within* $\left\lfloor \frac{D+1}{2} \right\rfloor$ *, where D is the diameter of the network.*

Proof. the proof is straightforward since the prescribed global equilibrium threshold ε is bounded by the diameter of the network D.

$\square$

We discuss the performance of introducing this rule in the next section.

3   Experimentation

In this section, we discuss some results through simulations that we conducted on NS2 (Network Simulator 2). We considered different sizes for the sensor network: 50, 100, 200, 400, 800 and 1600 nodes with an average density of 100 nodes per km². The radio transmission range is assumed to be 250 m. The threshold σ is set to 0.1 while $Max(0) - Min(0)$ is set to 10. The scalars of nodes and nodes positions are determined according to uniform distribution. We ran the two versions of our algorithm. The first version executes only the rule R_1 and the second executes both rules R_1 and R_2. For every size of the network, we consider 10 executions of the algorithm then we calculate the average of obtained results.

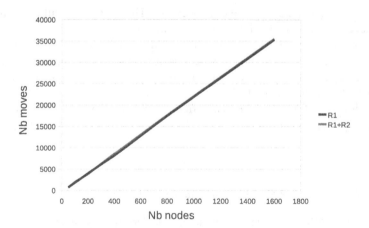

Fig. 4. Convergence time

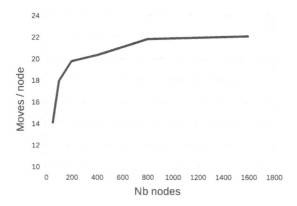

Fig. 5. Number of moves per node

We first discuss the convergence time. In our study we express this time by the number of moves performed by the set of all nodes of the network.

Consider the Figure 4. We can observe that both versions of the algorithms converge within similar amount of moves. We also observe that the number of moves is increasing linearly with the number of nodes in the network n. In fact, the slope of the line is about 22 while the one determined by the upper bound of the convergence time is exactly $(Max(0) - Min(0)) / \sigma = 100$. Actually, the equation of the obtained line is $y = 22.33\,x - 415$. Hence, we can expect that for large values of n, the number of moves performed by a node will be about 22 moves. This is confirmed by Figure 5.

This figure gives the number of moves performed by a node according to the total number of nodes in the network. The observed value increases in a logarithmic way until reaching the value of ~ 22.

Now, in order to show the performance of the introduction of the rule R_2, we consider the number of nodes that converge outside the interval $\overline{x} \pm \sigma$ where $\overline{x}$ is the global consensus average. Before giving interest to that number of nodes, we first discuss the ratio between σ and ε. In all our simulations, we observed that for every node i, after the convergence, the value $|x_i - \overline{x}|$ is always less than $3 \times \epsilon$. Hence, if we suppose a uniform distribution of sensed values or scalars, one might have no need of a prior knowledge or estimation of the diameter of the network to set σ according to a precision ε.

The Figure 6 gives the number of nodes that converge outside the interval $\overline{x} \pm \sigma$ (but still all the nodes converge within the interval $\overline{x} \pm \varepsilon$).

We can observe that for networks with small number of nodes, the introduction of the rule R_2 has not much effect. However, with the increase of the node number, the difference between the two versions of the algorithm become more important. Moreover, after the convergence, the value $|x_i - \overline{x}|$ is always less than $2 \times \epsilon$ if we consider the algorithm using both rules R_1 and R_2.

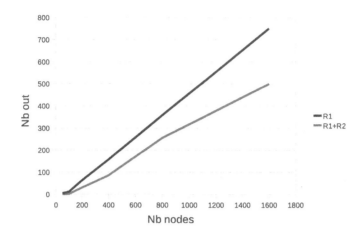

Fig. 6. Number of node out of $\bar{x} \pm \sigma$

4   Discussion and Future Work

We present in this section some generalizations of our algorithms. For the sake of simplicity, the discussion is given from the point of view of node i.

i) Improving Reliability: unexpected node failures may occur during the fusion process due to various reasons such as battery depletion/ exhaustion, software glitches, dislocation or environmental hazards and malicious attacks. To cope with this problem, when the area of interest has a significant density of sensors, we can perform redundancy/replication mechanisms, where some sensors can be in an active state: they participate in the network functioning while the others in a passive state (standby). These sensors wakeup periodically. If a working sensor node fails, it must be replaced by a passive one. However, two questions arise here: i) How the fault detection is done? and ii) how to replace the failed sensors?

These questions (i) and (ii) raise the following problems: (1) Since sensor nodes are not aware of their neighbors, especially the number of sleeping/passive nodes. How to adjust the wakeup period of these sensors? (2) During the recovery process, how to handle the case where two or more sleeping nodes, would realize at the same time that the working/active node is down? Indeed, for the same covered area, it should not contain several working nodes simultaneously, which would distort the computation of the average consensus, the *self stabilizing* algorithm should be built on the fact that only one sensor node must be in the *active* state for each covered area. To compute node's sleeping wakeup rate, we can borrow the same principle to [5]. Intuitively, nodes are initially in the sleeping mode. Each node sleeps for an exponentially distributed time generated according to a probability density function (PDF) $f(t) = \lambda e^{-\lambda t}$, where λ is the probing rate of the sensor node and t denotes its sleeping time duration.

ii) Distributed Termination: the detection of the conjunction of local terminations, which is a stable property, is a non-trivial problem. In fact it covers two issues: (i) detect whether all sensor node states converge to the average of the initial measurements of the network even when sensor nodes are subject to failures and (ii) ensure that we have achieved the desired computations. Solving this problem in a distributed manner, allows each sensor to detect that it has done and all the nodes reach this coherent state. Thus, the objective here is to overlay the *self stabilizing* iterative fusion process, a control mechanism that can detect the conditions of termination/convergence.

5   Conclusion

In this paper, we have addressed the problem of distributed data fusion in wireless sensor networks. This is a very natural and important problem, as several objectives (convergence, performance) must be considered simultaneously to fulfill the requirements of the user application. To the best of our knowledge, the proposed algorithm is the first to address the upper bound of the number of moves/iterations required to achieve/ensure the convergence of node states to the average of the initial measurements of the network. In addition, we also showed that the load difference between any two sensor nodes in the network is within $\frac{\varepsilon}{D} \times \left\lfloor \frac{D+1}{2} \right\rfloor < \varepsilon$, where ε is the prescribed global equilibrium threshold (this threshold is given by the system) and D is the diameter of the network.

Our approach should be extended to the context of safety critical applications. For instance, security threats must be addressed during the self-stabilizing fusion process. Most current approaches do not consider/include security measures, which opens an opportunity for further research in this field.

Acknowledgments. We thank the referees for all the valuable comments that helped us to improve the paper.

References

[1] Akyildiz, I., Su, W., Sankarasubramniam, Y., Cayirci, E.: A survey on sensor networks. IEEE Communications Magazine, 102–114 (2002)

[2] Paradis, L., Han, Q.: A survey of fault management in wireless sensor networks. JNSM 15(2), 171–190 (2007)

[3] Hai, L., Amiya, N., Ivan, S.: Fault-tolerant algorithms/protocols in wireless sensor networks. In: Handbook of Wireless Ad Hoc and Sensor Net., pp. 265–295 (2009)

[4] Saleh, I., Eltoweissy, M., Agbaria, A., El-Sayed, H.: A fault tolerance management framework for wireless sensor networks. JCM 2(4), 38–48 (2007)

[5] Ye, F., Zhang, H., Lu, S., Zhang, L., Hou, J.C.: A randomized energy-conservation protocol for resilient sensor networks. Wireless Networks 12(5), 637–652 (2006)

[6] de Souza, L.M.S., Vogt, H., Beigel, M.: A survey on fault tolerance in wireless sensor networks. Sap research, braunschweig, germany

[7] Lynch, N.: Distributed algorithms. Morgan Kaufmann Publishers, Inc. (1996)

[8] Cedo, F., Cortés, A., Ripoll, A., Senar, M.A., Luque, E.: The convergence of realistic distributed load-balancing algorithms. Theory Comput. Syst. 41(4), 609–618 (2007)

[9] Rabani, Y., Sinclair, A., Wanka, R.: Local divergence of markov chains and the analysis of iterative load-balancing schemes. In: Proceedings of the IEEE Symp. on Found. of Comp. Sci., Palo Alto (1998)

[10] Bahi, J., Couturier, R., Vernier, F.: Synchronous distributed load balancing on dynamic networks. Journal of Parallel and Distributed Computing 65(11), 1397–1405 (2005)

[11] Olfati-Saber, R., Murray, R.M.: Consensus problems in networks of agents with switching topology and time-delays. IEEE Transaction on Automatic Control 49(9), 1520–1533

[12] Bliman, P., Ferrari-Trecate, G.: Average consensus problems in networks of agents with delayed communications. Journal of IFAC 44(8), 1985–1995 (2008)

[13] Moallemi, C.C., Roy, B.V.: Consensus propagation. IEEE Trans. Inf. Theory 52(11), 4753–4766 (2006)

[14] Legg, J.A.: Tracking and sensor fusion issues in the tactical land environement. Technical Report TN.0605 (2005)

[15] Olfati-Saber, R., Shamma, J.S.: Consensus filters for sensor networks and distributed sensor fusion. In: 44th IEEE Conf. on Dec. and Cont. CDC-ECC (2005)

[16] Olfati-Saber, R.: Distributed kalman filter with embeded consensus filters. In: 44th IEEE Conf. on Dec. and Cont. (2005)

[17] Olfati-Saber, R., Fax, J., Murray, R.: Consensus and cooperation in networked multi-agent systems. In: Proc. of IEEE, pp. 215–233 (2007)

[18] Xiao, L., Boyd, S., Lall, S.: A space-time diffusion scheme for peer-to-peer least-squares estimation. In: Proc. of Fifth International Conf. on Information Processing in Sensor Networks (IPSN 2006), pp. 168–176 (2006)

[19] Talebi, M.S., Kefayati, M., Khalaj, B.H., Rabiee, H.R.: Adaptive consensus averaging for information fusion over sensor networks. In: IEEE International Conference on Mobile Adhoc and Sensor Systems (MASS), pp. 562–565 (2006)

[20] Kar, S., Moura, J.M.F.: Distributed consensus algorithms in sensor networks with imperfect communication: link failures and channel noise. IEEE Transactions on Signal Processing 57(1), 355–369 (2009)

[21] Bahi, J.M., Giersch, A., Makhoul, A.: A scalable fault tolerant diffusion scheme for data fusion in sensor networks. In: InfoScale 2008, pp. 1–5. ICST press (2008)

[22] Bertsekas, D.P., Tsitsiklis, J.N.: Parallel and Distributed Computation: Numerical Methods. Athena Scientific (1997)

[23] Gupta, S.K.S., Srimani, P.K.: Self-stabilizing multicast protocols for ad hoc networks. Journal of Parallel and Distributed Computing 63(1), 87–96 (2003); Wireless and Mobile Ad Hoc Networking and Computing

[24] Beauquier, J., Clement, J., Messika, S., Rosaz, L., Rozoy, B.: Self-stabilizing counting in mobile sensor networks. In: PODC 2007: Proceedings of the Twenty-Sixth Annual ACM Symposium on Principles of Distributed Computing, pp. 396–397. ACM, New York (2007)

[25] Hoepman, J.-H., Larsson, A., Schiller, E.M., Tsigas, P.: Secure and self-stabilizing clock synchronization in sensor networks. In: Masuzawa, T., Tixeuil, S. (eds.) SSS 2007. LNCS, vol. 4838, pp. 340–356. Springer, Heidelberg (2007)

[26] Dijkstra, E.W.: Self-stabilizing systems in spite of distributed control. Commun. ACM 17(11), 643–644 (1974)

[27] Gradinariu, M., Tixeuil, S.: Conflict managers for self-stabilization without fairness assumption. In: International Conference on Distributed Computing Systems, p. 46 (2007)

[28] Goddard, W., Hedetniemi, S.T., Jacobs, D.P., Srimani, P.K.: Self-stabilizing protocols for maximal matching and maximal independent sets for ad hoc networks. In: Proceedings of the 17th International Symposium on Parallel and Distributed Processing, IPDPS 2003, pp. 162.2. IEEE Computer Society, Washington, DC (2003)

[29] Goddard, W., Hedetniemi, S.T., Jacobs, D.P., Trevisan, V.: Distance- k knowledge in self-stabilizing algorithms. Theoretical Computer Science 399(1-2), 118–127 (2008); Flocchini, P., Gąsieniec, L. (eds.): SIROCCO 2006. LNCS, vol. 4056. Springer, Heidelberg (2006)

[30] Beauquier, J., Datta, A.K., Gradinariu, M., Magniette, F.: Self-stabilizing local mutual exclusion and daemon refinement. In: Herlihy, M.P. (ed.) DISC 2000. LNCS, vol. 1914, pp. 223–237. Springer, Heidelberg (2000)

[31] Afek, Y., Dolev, S.: Local stabilizer. Journal of Parallel and Distributed Computing 62(5), 745–765 (2002)

[32] Leal, W., Arora, A.: Scalable self-stabilization via composition. In: Proceedings of the 24th International Conference on Distributed Computing Systems (ICDCS 2004), pp. 12–21. IEEE Computer Society, Washington, DC (2004)

[33] Jaworski, J., Ren, M., Rybarczyk, K.: Random key predistribution for wireless sensor networks using deployment knowledge. Computing 85(1-2) (2009)

Recoverable Encryption through a Noised Secret over a Large Cloud

Sushil Jajodia[1], Witold Litwin[2], and Thomas Schwarz SJ[3]

[1] George Mason University, Fairfax, Virginia, USA
jajodia@gmu.edu
[2] LAMSADE, Université Paris Dauphine, Paris, France
witold.litwin@dauphine.fr
[3] Universidad Católica del Uruguay, Montevideo, Uruguay
tschwarz@ucu.edu.uy

Abstract. The safety of keys is the Achilles' heel of cryptography. A key backup at an escrow service lowers the risk of loosing the key, but increases the danger of key disclosure. We propose *Recoverable Encryption* (RE) schemes that alleviate the dilemma. RE encrypts a backup of the key in a manner that restricts practical recovery by an escrow service to one using a large cloud. For example, a cloud with ten thousand nodes could recover a key in at most 10 minutes with an average recovery time of five minutes. A recovery attempt at the escrow agency, using a small cluster, would require seventy days with an average of thirty five days. Large clouds have become available even to private persons, but their pay-for-use structure makes their use for illegal purposes too dangerous. We show the feaibility of two RE schemes and give conditions for their deployment.

Keywords: Cloud Computing, Recoverable Encryption, Key Escrow, Privacy.

1 Introduction

Data confidentiality ranks high among user needs and is usually achieved using high quality encryption. But what the user of cryptography gains in confidentiality he looses in data safety because the loss of the encryption key destroys access to the user's data. A frequent cause for key loss is some personal catastrophy that befalls the owner of the owner such as a fire that destroys the device(s) with passwords and keys. Organizations have to prevent a scenario where the sole employee with access to an important key leaves the organization or becomes incapacitated. In the past, keys were lost in natural disasters, such as when the basement of a large insurance (!) company was flooded and keys and their backups were destroyed. Many patients encrypt files with health data, but access to them becomes crucial especially if a health issue discapacitates the patient. Encrypted family data needs to be able to allow for their owner's disappearance, e.g. on a hiking trip in the Alaskan wilderness.

A. Hameurlain et al. (Eds.): TLDKS IX, LNCS 7980, pp. 42–64, 2013.

A common approach to key safety is a remote backup copy with some *escrow* service [JLS10]. The escrow service can be a dedicated commercial service, an administrator at the organization, a volunteer service, ... However, if the user entrusts a key to an escrow service, the user has to be able to trust the escrow service itself. The escrow service not only needs to prevent accidental and malicious disclosure by insiders and outsiders, but most be able to convince its users that the measures taken to protect the keys against disclosure or loss are of sufficient strength. This might explain why the use of escrow service is not very popular. No wonder that some users prefer to forego encryption [MLFR02], or prefer less security by using a weak or repeated password.

The *Recoverable Encryption* (RE) scheme [JLS10] is intended to alleviate the problem. It encrypts a backup of the key so that the decryption of the backup is possible without the owner using *brute-force*. Legitimate, authorized recovery is easy, while unauthorized recovery is computationally involved and dangerous. RE [JLS10] or Clasas [SL10] was designed for client-side encrypted data stored in a large LH* file in a cloud. The key backup is subjected to secret sharing and the shares are spread randomly over many LH* buckets. To recover the key, an adversary has to intrude buckets, which are stored in different sites, and search them until she can recover all or sufficient shares of the key. Thus, unauthorized recovery of the key involves illegal activity and is at best cumbersome.

Here, we present more general schemes, which we call collectively Recoverable Encryption through a Noised Secret (RENS). In these schemes, a single computer or cloud node suffices as the storage for key backups. The client uses an encryption of the backup key that resists brute force at the site of the escrow agency, decryption is possible through brute force by distributing the workload over the many nodes of a cloud. The user can choose an encryption strenght based on the maximum time D needed to recover the key at the site of the escrow service. On average, the time to key retrieval is $D/2$. The user will choose a time D in years or at least months, depending on his trust into the additional security measures of the escrow service. The user will also specify at maximum time R for legitimate recovery, which is in the range of minutes or even seconds. The relationship between R and D is given by the number N of cloud nodes needed for legitimate recovery and is approximately given by

$$N = D/R$$

Cloud computing has brought large-scale distributed computing to the masses. The possibility to rent a large number of standardized virtual servers for short amounts of time and that allow remote access changes the possibilities of a small organization or even an individual, bringing them tremendous compute power without investing into a proper IT infrastructure. The fact that this is a paid-for service brings additional benefits, as a cloud user can be forensically connected to any services used not only by user data and login information, but also by the money trail.

Nowadays, the number N of nodes is at least in the tens of thousands. Large clouds are now available to legitimate users. Today, (2012), Google and Yahoo claim to use clouds with more than hundred of thousands nodes and Microsoft's Azure advertizes millions of nodes. An unauthorized recovery of the key is clearly possible with these resources, but the cloud service providers are well aware of the potential of their resources for criminal activity and protect themselves against this possibility. Additionally, using legitimate clouds leaves many traces behind that can be used to trace and convict an adversary.

A legitimate recovery needs to rent the resources of such a large cloud and is somewhat costly. The amount depends of course on D, R, and the rental costs of the cloud. The user chooses R according to a trade-off between the urgency of an eventual recovery and the costs. An example that we later discuss shows that a public cloud with 8000 nodes would cost about a couple of hundred dollars. These costs are by themselves a deterrent to an escrow service who wants to "precompute their users' need". An escrow service would certainly not spend these amounts of money on recovering all keys, but when reimbursed by the user, be willing to broker the recovery using a cloud service. We can speculate that giving an economic cost to key recovery would make "key insurance" possible. The user might then protect herself against key loss by buying insurance at a nominal cost.

Technically, an RENS scheme hides the key within a *noised* secret. Like the classical shared secret scheme by Shamir [Sha79], a noised secret consits of two shares at least and the secret is the exclusive-or (xor) of the shares. At least one of the shares is *"noised"*, which means that it is hidden in a large set – the *noise space*. The size M of the noise space is a parameter set by the user who in this way determines D and R, both linearly proportional to M. With overwhelming probability only the noised share reveals the secret. The RENS recovery procedure searches for the noised share through the noise space by brute force. It recognizes the noise share because it is given a secure hash of the noised share as a *hint*. Decryption by searching for the true share within the noise space might need to inspect all shares, and will on average be successful after inspecting half of them.

If we move the search to the cloud, we can speed it up by parallelizing it. Two schemes are possible. We can use a *static* scheme where the number of nodes is selected before the search begins. A *scalable* scheme changes the number of servers if necessary in order to meet the deadline. If the through-put of each server is the same, then a static scheme will achieve the smallest cloud size N. Otherwise, a scalable scheme needs to be used.

A static scheme with a cloud of 10,000 nodes provides a speed-up that changes seconds into days, as a day has 86,400 seconds and minutes into months as a 30-day month has 43,200 minutes.

We use classical secret sharing to prevent any information leakage through the use of the cloud. Only one share of the key's backup is ever recovered by the cloud and the other share is retained by the escrow service. An adversary needs to gain access to both shares in order to obtain the key.

In the rest of this article, we analyze the feasibility of RENS schemes. We define the schemes, discuss correctness, safety, and the properties that we just outlined. We discuss related work in Section 2. Section 3 introduces the basic RENS scheme formally. The basic scheme assumes that the capacities of the nodes are approximately identical. We present a static scheme where the escrow service knows the capabilities of the nodes in advance. For instance, the escrow service rents hardware nodes from a cloud provider for a certain time. We then present a scheme that uses scalable partitioning where the nodes autonomously adjust their number to the task at hand. We present an optimization of this scheme that uses data from one additional node in order to lower the total number of nodes involved and hence the costs to the escrow server. We discuss the performance using simulation of an inhomogeneous cloud in Section 4. Our schemes so far does not give any assurance against finishing recovery early. We provide another extension in Section 5 to our basic idea that gives tight assurance for boundaries of the recovery time. For instance, we can guarantee with three nines assurance that the actual recovery time is between $1/2$ and 1 of the maximum recovery time. At the end, we conclude and discuss future work.

2   Related Work

The risks of key escrow are hardly a new issue. Key escrow mandated by government was a hotly contested issue in the nineties in the United States. Much work has been devoted to define the legal, ethical, and technical issues and to design, prototype, and standardize key recovery mechanisms. The work by Bellare and Goldwasser [BG97], the work on the Clipper proposal by US government [MA96] [Bla11], the proposal by Verheul and van Tilborg [VvT97], and the risk evaluation by Abelson and colleagues [AAB+97] on the technical side, and the ethical and legal assessments by Denning and Baugh [DBJ96] and Singhal [Sin95] among many others show this interest. The concept of recoverable encryption was implicit in Denning's taxonomy [DB96] and became more explict in a revised version [DB97]. Of course, we are considering here voluntary key escrow so that much of this work and criticism simply does not apply.

Gennaro and colleagues describe a two-phase key recovery system that allows reusing a single asymetrical cryptography operation to generate key recovery data for various sessions and give it to a recovery agent [GKM+97]. Ando *et al.* exhibit a method that replaces a human recovery agent with an automatic one [AMK+01]. Johnson and colleages patented a key recovery scheme that is interoperable with existing systems [JKKJ+00]. Gupta provides interoperability by defining a common key recovery block [Gup00], a work extended by Chandersekaran and colleagues, who patented a method for achieving interoperability between key recovery enabled and unaware systems [CG02], [CMMV05]. Andrews, Huang, and Ruan distribute information in order to simplify access to private keys in a public key infrastructure without sacrificing security [AHR+05]. D'Souza and Pandey allow data to be stored in a cloud system where the data store can release encrypted data upon receiving a threshold number of requests from third parties. The scheme is based on verifiable secret sharing [DP12]. Fan *et al.* give an overview of the state of the art [FZZ12]. Current work on key escrow in the scientific literature tries to avoid an unintended form of key escrow, where a public key generation system can reconstruct a client key [CS11].

We published the original Recoverable Encryption (RE) idea in 2010 [JLS10], where we applied it to data that a client encrypted and entrusted to the cloud. These data form an LH*$_{RE}$ file distributed over the nodes in the cloud. As its name suggests, this is a Linear Hash (LH) based Scalable Distributed Data Structure [LNS96], [AMR+11]. The encryption key was maintained by the user but also backed up in the cloud structure itself. The backup is subjected to secret sharing and to recover it, one has to collect all the shares. An authorized client of the cloud can use the LH*$_{RE}$ scan operation, but an intruder would have to break into typically many cloud nodes [JLS10], [SL10]. Whereas an LH*$_{RE}$ backup key is stored in the cloud itself, RENS only uses the cloud for the recovery itself.

In CSCP [LJS11], we also store files encrypted by the client in the cloud, but in contrast to LH*$_{RE}$ several users share keys among authorized clients. CSCP uses a static Diffie-Helman (DH) scheme. If a client looses her Diffie-Helman number, access to keys and files are lost, but an administrator has a backup of each private Diffie-Helman key. Obviously, RENS blends nicely with CSCP.

Our current proposal replaces the dispersion of the key into shares by a recovery scheme based on a targeted amount of computation. Whereas in previous schemes, the key was dispersed into a reasonably large number of shares, here, we only use two shares and allow access to one share through a limited computational effort. This concept has been made possible by the advent of "cloud computing" that puts large-scale distributed computing at the fingertips of the masses.

The concept of RE is rooted in the cryptographical concepts of *one-way hashes with trapdoor* and *cryptograms* or *crypto-puzzles* [DN93], [Cha11], [KRS+12]. RE can be considered to be a one-way hash where the computational capacity of cloud services for a distributed brute-force attack constitutes the trapdoor. RE in this sense is similar to Rivest's and Shamir's timed release crypto [RSW96], where a certain amount of computation needs to be performed in order to obtain a secret.

3   Recoverable Encryption through a Noised Secret

Recoverable encryption through a noised secret appears to the owner as the simple entrusting of the key in processed form to the escrow server, usually accompanied with some information for what the key is used. Upon request and after authentication and payment, the owner receives the key back from the escrow service after some processing time.

3.1   Client-Side Encryption

Before entrusting the backup of a key to the escrow service, the owner X preprocesses the key. The key is a bit-string of normal length (e.g. 256b for AES) that appears to be a random number.

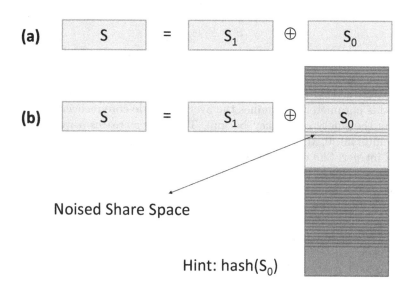

Fig. 1. Traditional secret sharing with two shares (a) and secret sharing with a noised secret (b)

```
import random

def create(S, M):
    S1 = random.getrandbits(KEYLENGTH)
    S0 = S1 xor S
    hashValue = hash(HASHALGO, S0)

    f = random.randint(0,M)
    l = int(S0) - f

    return S1, M, l, hashValue, HASHALGO

def recover(S1, M, l, hashValue, HASHALGO):
    for i in range(l, l+M):
        if hash(HASHALGO, i) == hashValue:
            S0 = i
    return S0 xor S1
```

Fig. 2. Pseudo-code for the creation of and the recovery from the noised secret

The owner uses classical secret sharing to write the key S as the exclusive or (xor) of two random strings of the same size as S:

$$S = S_1 \oplus S_0$$

The owner calculates the hash of S_0 using a standard, high-quality cryptographical hash method and stores $h(S_0)$ and a descriptor of the hash method as the *hint* $H(S)$ of the key. The owner chooses a size M of the noise space. As we will discuss, this parameter determines the average single-core recovery time D that represents the safety of the key backup. The owner creates a random number f in the interval $[0, M[$. The owner then converts S_0 from a bit string into an unsigned integer. She calculates $l = S_0 - f$. l forms the lower limit of the noise space that consists of the numbers l, $l+1$, ..., $l+M-1$. We call these numbers the *noise shares*, and refer to them collectively as the noise space. The true share S_0 is one of the noise shares and can be identified by the hint $h(S_0)$. Since we assume that the size of the hash is much larger than M, this is always possible with overwhelming probability. Figure 1 shows the procedure. The complete information given to the escrow service consists of S_1, M, l, and the hint $H(S)$.

We can still recover the original key S from this information. We iterate through the noise space starting with l and apply the hash to all noise shares. If we find one with the same hash as in the hint, we can assume that it is the true share S_0. We then recover the key as $S_1 \oplus S_0$. (Figure 2)

In order to protect against previously unknown vulnerabilities in the chosen hash method, we can choose an n-th power of a secure hash, i.e. calculate $h(S_0) = \phi^n(S_0)$ where ϕ is a NIST recommended standard hash function.

The owner uses the size M of the noise space in order to control the difficulty of the recovery operations. For this, she needs to have some reasonable estimate on the timing of the chosen hash function on a single-core processor together with a reasonable assumption on the number of cores that the escrow service or a bad employee of the escrow service might use. If she thinks that a reasonable number for the throughput of hash operations is T, then she obtains the maximum time D for recovery by the escrow using its own resources as

$$D = M/T$$

On average, an adversarial escrow service will use half that time to recover the noised share S_0 as the offset f of S_0 in the noise space was chosen randomly.

We need to be more carefull when we are using a private or public key created with one of the standard public key algorithms such as RSA, since the bits in such a key are highly redundant. It is known that an RSA key can be reconstructed from half of the bits [BM03, EJMDW05]. In case that we have a key that is not generated as a random bit string, we encrypt the key using a symmetric encryption method such as AES with a random key and then subject the latter key to our scheme. In this case, the usage information contains a description of the algorithm and the encryption of the original key.

3.2   Server Side Decryption

To recover a key, the escrow server has a share S_1, a size M of the noise space, and lower limit l of the noise space, and the hint $H(S)$, which contains the hash $h(S_0)$ of the noised share. The escrow server recovers the information using a brute-force attack, in which all elements of the noise space $l, l+1, \ldots, l+M-1$ are generated, their hash calculated, and compared with $h(S_0)$. With exceedingly high probability, there is only one share that has this hash value, namely the noised share S_0. The secret $S = S_0 \oplus S_1$ is returned.

The noise space is dimensioned so large that the server does not possibly have the means to perform this search with its own resources with any reasonable hope for success. It therefore needs to use a widely distributed computing service – a cloud service – in order to arme the recovery attempt. Brute force attacks are of course what is called "embarrassingly parallel" and can be easily partitioned into any number of sub-tasks that do not need to communicate amongst each other. If the server has Quality-of-Service (QoS) guarantees from the cloud provider, the easiest scheme is *static partitioning*, which we discuss first below. Otherwise, the server might use the principles of Scalable Distributed Data Structures (SDDS) [LNS96], (*scalable partitioning*), or a more involved interaction between a controller and participating working nodes. We describe a scalable partitioning scheme and two enhancements to deal with variations among node capacities below.

There is a (very) small chance for hash collisions, where there is more than one solution to $\text{hash}(X) = \text{hint}(= \text{hash}(S_0))$. A brute force attack will in general only returns the first solution found, which is not necessarily the true one. In this case, the escrow service will return a false key to the user. We assume that this becomes immediately obvious to the user who will complain to the escrow service. The escrow service will then repeat the search in an exhaustive manner, making sure to return all the possible solutions to $\text{hash}(X) = \text{hint}$. The probability of a collision is for a good hash close to the number of possible hashes divided by the size of the noise space. As good hashes have at least twenty bytes or one hundred and sixty bits, and as reasonable noise spaces do not have more than sixty bits, the chance for a hash collision is still in the order of 2^{-100} and probably much higher. If we want to protect against this already vanishingly small probability, we can do so at the costs of an additional hash. Since the changes necessary to switch to an exhaustive search are quite obvious, we do not consider this protection against the remote possibility of a hash collision in the following.

Example. A client wants to encrypt an AES key of length 512 bits. She wants D to be at least a month, i.e. 2^{22} seconds. She wants to be able to recover a key in minutes, leading her to set $R = 2^9$ seconds. Assume now that a node can make 2^{20} hash calculations per second. These numbers are reasonable in 2012 for a 2 GHz core processor, if we use SHA-256 as the hash. This gives us a noise space of 2^{20+22} or 2^{42} elements. Since the AES key is treated as an unsigned integer between 0 and 2^{512}, there is plenty of choice for the offset to an interval $I = [0, 2^{42}]$.

3.3   Decryption with Static Partitioning

If the server has guarantees for a minimum throughput of hash calculations at each node, the server determines the number of nodes necessary from the quality of service promise. If the maximal recovery time promised to the client is R, if a node can calculate at least T hashes per time unit, if N nodes are used, and if the size of the noise space is M then the ensemble can perform NT hashes per time unit. To evaluate a total of M hashes, it needs therefore M/NT time units, so that

$$\frac{M}{NT} \geq R$$

The minimum number of nodes needed is simply $M/(TR)$.

We can describe the algorithm using the popular map-reduce scheme. When the escrow server requests a cloud service, it deals directly only with one node, the *coordinator*. The coordinator calculates the number of *worker* nodes N. In the map-phase, the coordinator requests the N worker nodes and assigns them logical identification numbers 0, 1, ..., $N-1$. It also sends them the hint, the lower bound l of the noise space and the size M of the noise space.

In the reduce-phase, node a calculates the hash of the elements $l+a$, $l+a+N$, $l+a+2N$, ... and compares them with the hash of share S_0 contained in the hint. If it has found an element of the noise space with that hash, we assume that it has found S_0 and it sends a message with S_0 to the coordinator. If it has exhausted the search space, it sends a "terminated" message. Once the coordinator has received the result from one of the nodes, it will send a "stop" message to all nodes. A nodes that receives this message simply obeys.

In the termination phase, the coordinator sends the found string to the escrow server. This string is only one of the two shares, so the cloud itself has no information about the key. The escrow server now combines the two shares to obtain the key to return to the user.

Example Continued. Since $R = 2^9$ sec and $D = 2^{22}$ sec, $N = 2^{13} = 8192$. If we can rent a dedicated server core per hour at a cost of US$0.50 (November 2012), we would spend US$512.00 for an hour. If we can negotiate to pay for only part of the hour, the costs could sink to US$60.00 for the maximum time needed for recovery.

3.4   Recovery with Scalable Partitioning

For *scalable* or dynamic partitioning, we use the principles of Scalable Distributed Data Structures (SDDS) design to adjust the number of servers to the capabilities of the nodes. We assume that a node can reliably assess the throughput it can deliver for the time of the calculation. In order to distribute the work scalably and dynamically, *any* algorithm needs to make decisions based on the capabilities of only relatively few nodes. In this section, we present an algorithm where nodes make a decision on a split only based on their state. In the next section, Section 3.5, we provide two enhancements that use capacity information on the new node.

Our performance results (Section 4) show that they yield better performance measured in terms of the ratio of total capacity over total load. A smaller ratio means less nodes involved and hence less money paid to the cloud provider.

The scalable schemes go through the same phases as static partitioning. In the *initialization phase*, the escrow server selects a single cloud node (with index 0). The *map phase* immediately follows. Starting with the original node, each node compares its capacity with the task assigned to it and decides whether it needs to *split*, that is, request a new node from the cloud and share its workload with it. In the process of splits, each node acquires two parameters, its logical identifier and its level, that we use for the workload distribution. In this basic scheme, nodes only use local information in order to decide whether to split.

At the beginning of the mapping phase, Node 0 calculates its throughput capability B_0 given its current load. This throughput calculation is repeated at each node used in the recovery procedure. The node has a number n of hashes to calculate, a maximum time R to perform all of these calculations, and calculates a rate τ of calculations. A node then calculates its *load factor* $\alpha = \tau n / R$. If $\alpha > 1$, then the node is overloaded. If the initial node 0 has $\alpha \leq 1$, it is capable of doing the whole calculation itself, which it does and then returns the result to the escrow service. In the much more likely opposite case, Node 0 requests a new node from the cloud service provider, which becomes Node 1. The noise interval is divided into two equal halves and each half is assigned to one of the two nodes. Both new nodes acquire a new *level* $j = 1$.

Each node calculates its load factor α. If the load factor is larger than one (the node is overloaded), it *splits*. A split effectively divides the work assigned to the node between that node and a new node. Thus, each split operation requests a new node from the cloud server and incorporates it into the system. If node i with level j has split, then the node increases its level to $j + 1$, and the new node receives number $i + 2^j$ and level $j + 1$.

We recall that the noise space starts with number l. The node with identity number n and level j calculates the hashes $l + x$, where $x \equiv n \bmod 2^j$ and $0 \leq x < M$. LH* addressing [LNS96] guarantees that element in the noise space is assigned to exactly one node.

As in the static scheme, a node that finds a solution and therefore with overwhelming probability the noised share S_0 sends its find to Node 0. This constitutes the reduce phase. In the *termination phase*, Node 0 asks all other nodes to stop. It does so by sending the stop message to all nodes that split from it, i.e. to Nodes 1, 2, ,4, 8, …… Each node that receives the stop message, forwards its to all nodes that have split from it. The number of messages that a node has to send or forward is lmited by its level and therefore logarithmic in the number of nodes.

Example. We assume a very small example with nodes of largely varying capacity. Node 0 receives a workload of 15000 hashes and estimates that it can calculate 10000 hashes. Therefore, its workload factor α is 1.5 or 150%. It therefore splits. The new node has logical address $1 = 0 + 2^0$ and both nodes have level 1. Node 1 estimates that it can calculate 2000 hashes and has therefore a load factor $\alpha = 3.75$, while the load factor at Node 0 has been halved to .75. Node 0 therefore stops splitting, but Node 1 will have do, claiming a new node with logical address $3 = 1 + 2^1$. Node 3 decides that it can handel 11000 hashes and has therefore a load factor of 0.295, whereas Node 1 has a load factor of 1.875. Therefore, Node 1 splits once more, requesting a new Node with identity number $5 = 1 + 2^2$. Its load sinks to 1625 hashes and its new load factor is .8125. If the new node 5 can handle 9000 hashes, then its load factor is 0.181, so that there are no more splits.

We now have a total of four nodes. Node 0 has level 1, Node 1 has level 3, Node 3 has level 2, and Node 5 has level 3. Assume that $l = 1000000$, so that the noise space is $[1000000, 1015000[$. Node 0 calculates the hashes of all even numbers, i.e. 1000000, 1000002, 10000004, ..., 1014998, using an increment of 2^1, since it has level 1. Node 1 has level 3, therefore an increment of 8, and calculates 1000001, 1000009, 1000017, ... Node 3 has level 2 and therefore an increment of 4, so that it calculates 1000003, 1000007, ... Node 5 has level 3, an increment of 8, and calculates 1000005, 1000013, ...

3.5   Scalable Partitioning with Limited Load Balancing

To scale well, scalable partitioning needs to minimize the interchange of information between nodes. In real life instances, the load factor of the initial node is several tens of orders of magnitude larger. For example, a scheme where the coordinator polls potential nodes for their capacity in order to use an optimal assignment is completely out of the question. In the current scalable partitioning scheme, decision on splits are made based on information only at the level of a single node. A good solution will have to balance the speed of making decisions only at the local level with the overprovisioning caused by variations in the capacity of the nodes. In the previous example, the problems stem from Node 1, which has only one fifth of the capacity of the initial load. If Node 0 and Node 3 would have been used at their full capacity, the incorporation of Node 5 would have become superfluous.

Besides allowing limited communication between nodes, we also need to change to a more flexible assignment of load. We now use a type of range partitioning to assign loads. Now nodes calculate the hashes of a contiguous range of numbers $[x_0, x_1[$ of numbers within the noise interval $[l, l + M[$. If node p with an assignment of $[x_0, x_1[$ splits, it decides on a cutoff point p_1 and assigns itself the workload $[x_0, p_1[$ and to the new node the interval $[p_1, x_1[$. During the first phase of the map-phase, p_1 will be the midpoint $\lfloor (x_0 + x_1)/2 \rfloor$. Our enhancements have the splitting node use the capacity of the new node when calculating p_1.

We only investigate here two enhancements of the scheme where during a split the new node sends the information about its capacity to the splitting node. Our first strategy has the splitting node p detect if the capacity of the new node n and its own capacity suffice to perform the work assigned currently to p. For example, if node p has a capacity of 0.8 in order to do work 1.8, it has to split. If the new node has capacity 1.2, the combined capacity of 2.0 is sufficient to do the work. However, if we distribute the work equally, p will have work of 0.9

assigned to it, and will have to split again, whereas Node n has spare capacity. In the first improvement strategy, node p will get 0.8 work and n will get 1.0 work.

Our second, additional strategy has a node decide whether the load distribution is getting close to achieve its goal. If node p has a capacity c_p and a currently assigned workload of $w < 3 \cdot p$, it will split, but assign to itself only the work that is within its capacity. The new node is likely to have to split itself, but probably (though not for sure) no more than once. Our full enhancement uses both strategies, but can be obviously expanded by interchanging information between more nodes. We have to leave the exploration of these issues to future work.

Example (Continued). If we use the full enhancement in the previous example, then Node 0 communicates with Node 1 in order to obtain its capacity. Since the combined capacity of both nodes is 12000 and the total load is 15000, the first strategy is not employed. However, since the capacity of Node 0 is within "striking distance" of the load, it assigns itself 10000 hashes (the numbers in $[1000000, 1010000[$) and the remainder (the numbers in $[1010000, 1015000[$) to Node 1. The load factor of Node 1 after this split is $5000/2000 = 2.5$ and it still has to split. Since the capacity of the new node, Node 3, is 11000, the combined capacity of Nodes 1 and 3 is sufficient. Therefore, Node 1 splits its load at a ratio of $2 : 11$. It therefore assigns to itself the interval $[1010000, 1010769[$ and to Node 3 the interval $[1010769, 1015000[$. In this case, more extensive communication between Nodes 0, 1, and 3 could have let to a more balanced distribution, but not employed less nodes. The total capacity of the three nodes is 23000, so that we still overprovide. A more sophisticated scheme could have liberated Node 1, since its potential contribution is not only marginal, but also unnecessary.

4   Performance Analysis

Static partitioning always yields the best utilization of cloud nodes, but assumes that the throughputs at all nodes are perfectly even and known at the beginning of a run.

Scalable partitioning allows nodes to have different capacities, and detects these capacities whenever a new node is added. If nodes have all the same capacities, then a node will be split and its load divided by two until the load is less than 1 times the capacity of a node. If the total load is l times the node capacity, then Node 0 is split $\lfloor \log_2(l) \rfloor + 1$ times. This gives us the ratio of total capacity over total load to be

$$2^{\lfloor \log_2(l) \rfloor + 1}/l$$

This functions oscillates between 1.0 and 2.0 as Figure 3 shows. The average ratio is $\log(2) = 1.38629$ and is the price we pay for scalability.

If the capacity of the nodes is not constant but instead is subject to a non-constant probability distribution, then a different picture emerges. We assumed

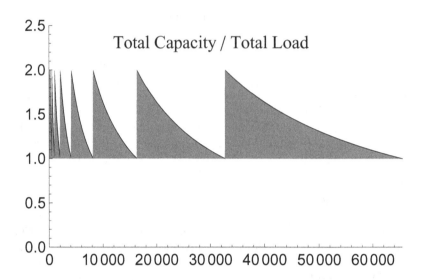

Fig. 3. Ratio of total capacity over total load with identical capacity at each node

first that the capacity of the node is normally distributed around l times the node capacity with different standard deviation and simulated the ratio. The simulation is accurate to three or four digital digits. The result of our simulation is given in Figures 4 and 5, where the standard deviation is 10%, 25%, 33%, and 50%.

The simple enhancement (as discussed previously) determines if a splitting node and the new node together have the capacity to perform the assigned task. In this case, the task is split according to capacity. Otherwise, the task is split evenly among the splitting and new node. In this case, at least one of them has to split.

The enhancement (as also discussed previously) includes the simple enhancement. If this is not the case, but if the assigned load is within three times its capacity, then the splitting node assign to itself all the load it can handle and passes the rest of the load to the new node. The assumption is that frequently, the new node will only have to split once.

We first observe that the basic variant now performs more consistently than without variation in the node capabilities. If the standard deviation is small, it exhibits overprovisioning close to the expected rate. However, the ratio of total capacity over total load for the basic scalable partitioning scheme increases with increased standard deviation. For 50% standard deviation, its ratio is consistently higher than 2. (In our simulation, we used a minimum capacity of $1/100$ so that the probability distribution is strictly speaking no longer normally distributed, which would allow for negative capacities. As the standard deviation increases, the mean capacity therefore slightly increases as well.) With increasing deviation, the oscillations become much less pronounced. The simple enhancement shows visible improvements with all standard deviations, but for 10% standard deviation only in the dips of the curve. The full enhancement

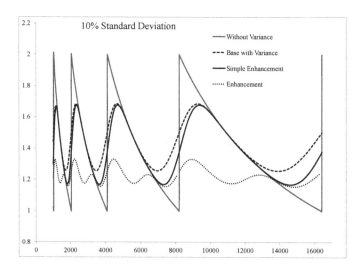

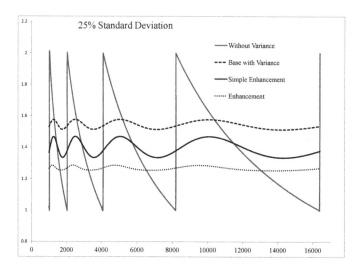

Fig. 4. Ratio of total capacity over total load depending on the load given in terms of expected node capability, using scalable partitioning without variation,scalable partitioning with normally distributed node capacity with standard deviations of 10% and 25% and with two of our extensions

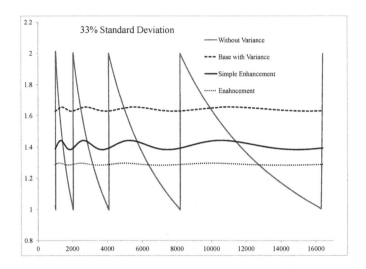

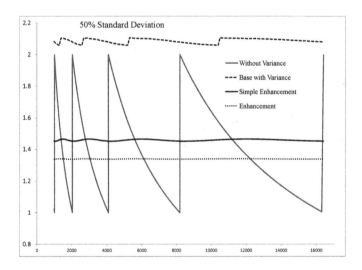

Fig. 5. Ratio of total capacity over total load depending on the load given in terms of expected node capability, using scalable partitioning without variation, scalable partitioning with normally distributed node capacity with standard deviations of 33% and 50% and with two of our extensions

Table 1. Average values of total capacity over total load ratios

Standard Deviation	Average Total Capacity over Total Load
Base with Variance	
10%	1.438
25%	1.542
33%	1.641
50%	2.086
Weibull 50%	1.856
Gamma 50%	2.170
Simple Extension	
10%	1.382
25%	1.393
33%	1.409
50%	1.458
Weibull 50%	1.478
Gamma 50%	1.598
Extension	
10%	1.219
25%	1.266
33%	1.289
50%	1.339
Weibull 50%	1.357
Gamma 50%	1.465

shows continuous improvements over the base and the simple enhancement. We notice however that the average increases slightly as is shown in Table 1.

When we simulated a scenario where the capacity of the nodes follows a different distribution, namely a gamma distribution with mean 1.0 and standard deviation of 50% and a Weibull distribution with the same mean and distribution, we found that the average ratio of total capacity over load was close to being constant, not depending on the total load. As was to be expected, the distribution is a major factor in the ratio. However, the benefits of the two extensions considered were equally obvious, though in the case of the Weibull distribution to a slightly lesser degree.

We show the effects of varying the standard deviation in Figures 6 and 7, which shows that the use of *local* capacity information when distributing small remaining load among few nodes is beneficial. The enhancements to the protocol do better in the case of the gamma distribution, since the gamma distribution (a convolution of the exponential distribution) has more small capacity nodes. We should note that our choice of probability distributions serve just as a stand-in for the unknown distribution. Much more research and measurements are necessary in this area.

The basic idea of exchanging information in the final phase of mapping does not violate the principles of scalability. In these scenarios, a node in the mapping phase enters a *final assignment* state whenever its assigned load is within c times its capacity, where c is a relatively small number. In this state, the node recruits

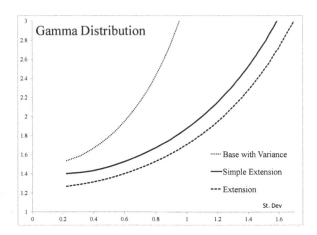

Fig. 6. Ratio of total capacity over total load in dependence on the standard deviation

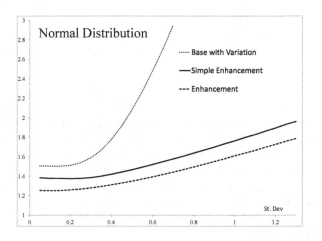

Fig. 7. Ratio of total capacity over total load in dependence on the standard deviation

new nodes one by one until there are enough nodes left to deal with the workload. In the worst case, this method leaves the last recruited node with only a marginal workload. In expectation, the number of nodes recruited would be between c and $c + 1$, so that we can estimate a reasonable upper bound on the load factor to be $1 + 1/c$.

5   Multiple Noises

In our scheme, the worst-time recovery at the esrow service is R, but the best possible time is negligible, since the very first hash calculated might yield the noised share. Some users averse to gambling might find this prospect discomforting. For this group, we present now a solution that gives guarantees against obtaining the backup of the key too quickly.

The chance to obtain the noised share within time ρR (where R is the maximum time) is equal to ρ. It is well known, that the last of n uniformly distributed tasks has a much smaller spread. In our case, this leads to *multiple noising*. There, we require the escrow service to use brute force in the cloud to invert n hashes.

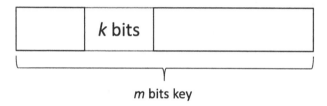

Fig. 8. Selection of noised part of key for multiple noises

Assume that we want a maximum of 2^k hashes to be calculated, that the key length is $m > k$, and that we want to invert n hashes. We recall that our scheme splits the key S into two different shares S_0 and S_1 of the same length and that share S_0 is being noised. We select k among the m bit positions in the key. Figure 8 shows a selection of k contiguous bits. The share S_0 is the concatenation of the selected bits and the $m - k$ remaining bits. We write this concatenation as $S_0 = I \sqcup R$, where I is made up of the k selected bits and R of the remaining bits. We now use classical secret sharing writing $I = I_1 \oplus I_2 \oplus \ldots I_n$, where the I-shares $I_1, I_2, \ldots I_n$ are random bit strings. We calculate the hashes $H_\nu = h(I_\nu \sqcup R)$ as the core part of n hints. (The remainder of the hints contains information about the hash selected and the length k of I and $m - k$ of R.)

For server side decryption, the escrow service uses a cloud to solve in parallel

$$h(X \sqcup R) = H_\nu \qquad \nu \in \{1, 2, \ldots, n\}$$

After it has found all solutions $J_1, J_2, \ldots J_n$, the share S_0 is calculated as

$$S_0 = (J_1 \oplus J_2 \oplus \ldots \oplus J_n)) \sqcup R$$

This calculation terminates after the last of the n equations has been solved.

The expected time to recover the key is the expected time to recover the last of the n shares $J_1, J_2, \ldots J_n$. We normalize the maximum recovery time to 1. Let the random variable X_i represent the time to recover share J_i. The

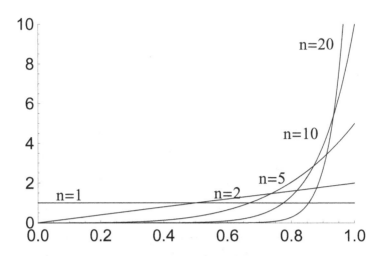

Fig. 9. Probability density for the maximum of m uniformly distributed random variables in $[0, 1]$

Table 2. Three and six nines guarantees that the last of n hashes is inverted in no less than p

n	p (three nines)	p (six nines)	Expected Value
1	0.001	10^{-6}	0.500
3	0.100	0.010	0.750
5	0.252	0.063	0.833
10	0.501	0.251	0.909
20	0.708	0.501	0.952

random variables are identically and independently distributed. The cumulative distribution function for the time to recover the last of the n shares of S_0 is then

$$F(x) = \text{Prob}(\max(X_i) < x) = \text{Prob}(X_1{}^n < x) = x^n$$

The probability density function of recovering all n shares is thus given by nt^{n-1} (Figure 9). Consequentially, the probability of key recovery in an exceedingly short time is made very low. The mean time to recovery is then $n/(n+1)$.

We can also give minimum time guarantee at a certain assurance level a such as $a = 0.999$ (three nines) or $a = 0.999999$ (six nines. This is defined to be the time value x_0 such that $\text{Prob}(\max(X_i) < x_0) = a$, i.e. that with probability (at least) a, the recovery takes more than x_0 times the maximum recovery time. Table 2 gives us the guarantees. For example, if we choose a safety of six nines, then we know at this level of assurance that the last of 20 shares will be recovered in less than 0.501 or 50% of the maximum recovery time.

6 Security

The security of our scheme is measured by the inability of an attacker to recover a key entrusted to the escrow service. An attacker outside of the escrow service needs to obtain both shares S_1 and S_0 of the key. This is only possible by breaking into the escrow server and becoming effectively an insider. We can therefore restricts ourselves to an attacker who is an insider. In this case, we have to assume that the attacker can break through internal defenses and obtain S_1 and the hint $h(S_0)$ of the noised secret. The insider then has to invert the hash function in order to obtain S_0.

We can systematically list the possibilities:

It is possible that there is no need to invert the hash function. As already mentioned in Section 3.1, RSA keys can be reconstructed from about half of the bits [BM03, EJMDW05]. If the scheme would be applied to keys that cannot be assumed to be random bits, then the specification of the noise space could be sufficient to generate a single or a moderate number of candidate keys just from the knowledge of the noise space. The insider attacker can then easily recover S_0 and therefore the key.

The inversion of the hash in the noise space could be much simpler than assumed. Cryptography is full of examples of more and more powerful attacks, as the history of MD5 and WEP show. In addition, the computational power of a single machine has increased exponentially at a high rate since the beginning of computing. The introduction of more powerful Graphical Processing Units (GPU) [OHL+08, KO12], has lead to a one-time jump in the capabilities of relatively cheap servers. It is certainly feasible that GPU computing can enter the world of for-hire cloud computing. Even if this is not the case, then competition, better energy use, and server development should lower the costs of computation steadily. This is a real problem for our scheme, but shares it with much of cryptographical methods. Just as for example key length has to be steadily increased, so the size of our noise spaces needs to be increased in order to maintain the same degree of security. Only, our system has to be more finely tuned as we cannot err on the side of exaggerated security. A developer worried about computational attack on a certain cryptographical scheme can always double the key size "just to be sure" and the product will only show a slight deterioration in response time due to the more involved cryptographical calculations. In our scheme, this is not an option. On the positive side, there is no new jump in sight that would increase single machine capability as the introduction of GPU computing did, and this one came with ample warning. Second, the times of the validity of Moore's law seem to be over, as single CPU performance cannot be further increased without incurring an intolerable energy use. The new form of Moore's law will be a steady decline in the decrease of the costs of single CPU computation. Overall, the managerial challenges of decreasing computation costs seem to be quite solvable.

Finally, the insider attacker could use the recovery scheme itself, availing herself of anonimity servers and untraceable credit cards, as are sold in many countries for use as gifts. This is a known problem for law enforcement as spammers can easily use the untraceability of credit cards in order to set up fraudulent websites. However, any commercial service that accepts this type of untraceable credit card opens itself up to charges of aiding and abetting and at least of gross negligence. When we are assessing these type of dangers, we need to be realistic. Technology such as cryptography only defines quasi-absolute security, but assuming a certain social ambience. If I want to read my boss's letters, I have certainly the technical tools to open an envelope (apparently hot water steam is sufficient), read the letter, and use a simple adhesive to close the letter. But even if I were inclined to do so, the social risk is inacceptable. In our case, an insider or an outside attacker that has penetrated the escrow service would have to undertake an additional step with a high likelihood of leaving traces. People concerned with security in organizations at high and continued risk know that adversaries usually resort to out-of-band methods. West-German ministries in the eighties were leaking secret information like sieves not because of technical faults but because of Romeo-attacks, specially trained STASI agents seducing well-placed secretaries.

7   Conclusions

We have introduced a new password recovery scheme based on an escrow service. Unlike other escrow based schemes, in our scheme the user knows that the escrow server will not peek at the data entrusted to it, as it would cost too much. Our scheme is based on a novel idea of using the scalable and affordable power of cloud computing as a back door for cryptography. Recoverable encryption could even be considered a new form of cryptography.

The relatively new paradigm of cloud computing still has to solve questions such as reliable quality of service guarantees and protection against node failures. In our setting, ignoring the issue is a reasonable strategy, since the only node that matters (*ex post facto*) is the one that will find the noised share. The expected behavior of recovery is hence the one of that single server and the quality of service of that single server is the one experienced by the end-user. However, our discussion on how to distribute an embarrassingly parallel workload in a cloud with nodes of varying capacity should apply to other problems. In this case, scalable fault-resilience does become an interesting issue. For instance, cloud virtual machines can suffer capacity fluctuations because of collocated virtual machines. We plan to investigate these issues in future work.

References

[AAB⁺97]   Abelson, H., Anderson, R., Bellovin, S.M., Benaloh, J., Blaze, M., Diffie, W., Gilmore, J., Neumann, P.G., Rivest, R.L., Schiller, J.I., Schneier, B.: The risks of key recovery, key escrow, and trusted third-party encryption. World Wide Web Journal 2(3), 241–257 (1997)

[AHR⁺05]   Andrews, R.F., Huang, Z., Ruan, T.Q.X., et al.: Method and system of securely escrowing private keys in a public key infrastructure. US Patent 6,931,133 (August 2005)

[AMK⁺01]   Ando, H., Morita, I., Kuroda, Y., Torii, N., Yamazaki, M., Miyauchi, H., Sako, K., Domyo, S., Tsuchiya, H., Kanno, S., et al.: Key recovery system. US Patent 6,185,308 (February 6, 2001)

[AMR⁺11]   Abiteboul, S., Manolescu, I., Rigaux, P., Rousset, M.C., Senellart, P.: Web data management. Cambridge University Press (2011)

[BG97]   Bellare, M., Goldwasser, S.: Verifiable partial key escrow. In: Proceedings of the 4th ACM Conference on Computer and Communications Security, pp. 78–91. ACM (1997)

[Bla11]   Blaze, M.: Key escrow from a safe distance: looking back at the clipper chip. In: Proceedings of the 27th Annual Computer Security Applications Conference, pp. 317–321. ACM (2011)

[BM03]   Blömer, J., May, A.: New partial key exposure attacks on RSA. In: Boneh, D. (ed.) CRYPTO 2003. LNCS, vol. 2729, pp. 27–43. Springer, Heidelberg (2003)

[CG02]   Chandersekaran, S., Gupta, S.: Framework-based cryptographic key recovery system. US Patent 6,335,972 (January 1, 2002)

[Cha11]   Chandrasekhar, S.: Construction of Efficient Authentication Schemes Using Trapdoor Hash Functions. PhD thesis, University of Kentucky (2011)

[CMMV05]   Chandersekaran, S., Malik, S., Muresan, M., Vasudevan, N.: Apparatus, method, and computer program product for achieving interoperability between cryptographic key recovery enabled and unaware systems. US Patent 6,877,092 (April 5, 2005)

[CS11]   Chatterjee, S., Sarkar, P.: Avoiding key escrow. In: Identity-Based Encryption, pp. 155–161. Springer (2011)

[DB96]   Denning, D.E., Branstad, D.K.: A taxonomy for key escrow encryption systems. Communications of the ACM 39(3), 35 (1996)

[DB97]   Denning, D.E., Branstad, D.K.: A taxonomy for key escrow encryption systems (1997), faculty.nps.edu/dedennin/publications/TaxonomyKeyRecovery.htm

[DBJ96]   Denning, D.E., Baugh Jr., W.E.: Key escrow encryption policies and technologies. Villanova Law Review 41, 289 (1996)

[DN93]   Dwork, C., Naor, M.: Pricing via processing or combatting junk mail. In: Brickell, E.F. (ed.) CRYPTO 1992. LNCS, vol. 740, pp. 139–147. Springer, Heidelberg (1993)

[DP12]   D'Souza, R.P., Pandey, O.: Cloud key escrow system. US Patent 20,120,321,086 (December 20, 2012)

[EJMDW05]   Ernst, M., Jochemsz, E., May, A., de Weger, B.: Partial key exposure attacks on RSA up to full size exponents. In: Cramer, R. (ed.) EUROCRYPT 2005. LNCS, vol. 3494, pp. 371–386. Springer, Heidelberg (2005)

[FZZ12]   Fan, Q., Zhang, M., Zhang, Y.: Key escrow attack risk and preventive measures. Research Journal of Applied Sciences 4 (2012)

[GKM⁺97]   Gennaro, R., Karger, P., Matyas, S., Peyravian, M., Roginsky, A., Safford, D., Willett, M., Zunic, N.: Two-phase cryptographic key recovery system. Computers & Security 16(6), 481–506 (1997)

[Gup00]   Gupta, S.: A common key recovery block format: Promoting interoperability between dissimilar key recovery mechanisms. Computers & Security 19(1), 41–47 (2000)

[JKKJ⁺00]   Johnson, D.B., Karger, P.A., Kaufman Jr., C.W., Matyas Jr., S.M., Safford, D.R., Yung, M.M., Zunic, N.: Interoperable cryptographic key recovery system with verification by comparison. US Patent 6,052,469 (April 18, 2000)

[JLS10]   Jajodia, S., Litwin, W., Schwarz, T.: LH*RE: A scalable distributed data structure with recoverable encryption. In: CLOUD 2010: Proceedings of the 2010 IEEE 3rd International Conference on Cloud Computing, pp. 354–361. IEEE Computer Society, Washington, DC (2010)

[KO12]   Komura, Y., Okabe, Y.: Gpu-based single-cluster algorithm for the simulation of the ising model. Journal of Computational Physics 231(4), 1209–1215 (2012)

[KRS⁺12]   Kuppusamy, L., Rangasamy, J., Stebila, D., Boyd, C., Nieto, J.G.: Practical client puzzles in the standard model. In: Proceedings of the 7th ACM Symposium on Information, Computer and Communications Security, ASIACCS 2012. ACM, New York (2012)

[LJS11]   Litwin, W., Jajodia, S., Schwarz, T.: Privacy of data outsourced to a cloud for selected readers through client-side encryption. In: WPES 2011: Proceedings of the 10th Annual ACM Workshop on Privacy in the Electronic Society, pp. 171–176. ACM, New York (2011)

[LNS96]   Litwin, W., Neimat, M.A., Schneider, D.A.: Lh* – a scalable, distributed data structure. ACM Transactions on Database Systems (TODS) 21(4), 480–525 (1996)

[MA96]   McConnell, B.W., Appel, E.J.: Enabling privacy, commerce, security and public safety in the global information infrastructure. Office of Management and Budget, Interagency Working Group on Cryptography Policy, Washington, DC (1996)

[MLFR02]   Miller, E.L., Long, D.D.E., Freeman, W.E., Reed, B.C.: Strong security for network-attached storage. In: Proceedings of the 1st USENIX Conference on File and Storage Technologies, p. 1. USENIX Association (2002)

[OHL⁺08]   Owens, J.D., Houston, M., Luebke, D., Green, S., Stone, J.E., Phillips, J.C.: Gpu computing. Proceedings of the IEEE 96(5), 879–899 (2008)

[RSW96]   Rivest, R.L., Shamir, A., Wagner, D.A.: Time-lock puzzles and timed-release crypto. Technical report, Massachusetts Institute of Technology, Cambridge, MA, USA (1996)

[Sha79]   Shamir, A.: How to share a secret. Communications of the ACM 22(11), 612–613 (1979)

[Sin95]   Singhal, A.: The piracy of privacy-a fourth amendment analysis of key escrow cryptography. Stanford Law and Policy Review 7, 189 (1995)

[SL10]   Schwarz, T., Long, D.D.E.: Clasas: a key-store for the cloud. In: 2010 IEEE International Symposium on Modeling, Analysis & Simulation of Computer and Telecommunication Systems (MASCOTS), pp. 267–276. IEEE (2010)

[VvT97]   Verheul, E.R., van Tilborg, H.C.A.: Binding ElGamal: A fraud-detectable alternative to key-escrow proposals. In: Fumy, W. (ed.) EUROCRYPT 1997. LNCS, vol. 1233, pp. 119–133. Springer, Heidelberg (1997)

Conservative Type Extensions for XML Data

Jacques Chabin[1], Mirian Halfeld Ferrari[1],
Martin A. Musicante[2], and Pierre Réty[1]

[1] Université d'Orléans, LIFO, Orléans, France
[2] Universidade Federal do Rio Grande do Norte, DIMAp Natal, Brazil

Abstract. We introduce a method for building a *minimal* XML type (belonging to standard class of regular tree grammars) as an extension of other given types. Not only do we propose an easy-to-handle XML type evolution method, but we prove that this method computes the smallest extension of a given tree grammar, respecting pre-established constraints. We also adapt our technique to an interactive context, where an advised user is guided to build a new XML type from existing ones. A basic prototype of our tool is implemented.

1 Introduction

We deal with the problem of exchanging valid XML data in a multi-system environment. We assume that $I_1, \ldots, I_n$ are local systems that inter-operate with a global system I which should be capable of receiving information from any local system. Local systems $I_1, \ldots, I_n$ deal with sets of XML documents $X_1, \ldots, X_n$, respectively. Each set X_i conforms to schema constraints $\mathcal{D}_i$ and follows an ontology O_i. We want to associate I to a schema for which documents from any local system are valid, and, in this way, we consider that this new schema $\mathcal{D}$ is an evolution for all local systems.

Every real application goes through type evolution, and, thus, in general, our approach is useful whenever one wants not only to generate XML types from given ones but also to preserve the possibility of processing previous versions of software systems. In other words, we focus on a conservative type evolution, *i.e.*, in allowing some backward-compatibility on the types of data processed by new versions, in order to ensure that old clients can still be served. In a multi-system environment, this means to keep a service capable of processing information from any local source, without abolishing type verification.

In the XML world, it is well known that type (or schema) definitions and regular tree grammars are similar notions and that some schema definition languages can be represented by using specific classes of regular tree grammars. As mentioned in [Mani and Lee, 2002], the theory of regular tree grammars provides an excellent framework for understanding various aspects of XML type languages. They have been actively used in many applications such as: XML document processing (*e.g.* XQuery and XDuce)[1] and XML document validation

---

[1] http://www.w3.org/TR/xquery/ and http://xduce.sourceforge.net/

A. Hameurlain et al. (Eds.): TLDKS IX, LNCS 7980, pp. 65–94, 2013.
© Springer-Verlag Berlin Heidelberg 2013

algorithms (*e.g.* RELAX-NG[2]). They are also used for analysing the expressive power of the different schema languages [Murata et al., 2005].

Regular tree grammars define sets of trees (in contrast to more traditional string grammars, which generate sets of strings). The rules of *general regular tree grammars* (or RTG) have the form $X \to a\,[R]$, where X in a non-terminal symbol, a is a terminal symbol, and R is a regular expression over non-terminal symbols. *Local tree grammars* (LTG) are regular tree grammars where rules for the same terminal symbol have the same non-terminal on their left-hand side. *Single-type tree grammars* (STTG) are regular tree grammars where distinct non-terminal symbols which appear in a same regular expression of the grammar, always generate distinct terminal symbols. Notice that the restriction for a grammar to be an LTG is stronger than the restriction for STTGs. This means that every LTG is also an STTG.

The interest of regular tree grammars in the context of XML progressing is that each of the two main stream languages for typing XML documents, *i.e.*, DTD[3] and XML Schema (XSD)[4], correspond, respectively, to LTG and STTG [Murata et al., 2005].

Given an XML type and its corresponding tree grammar G, the set of XML documents described by G corresponds to the language (set of trees) $L(G)$ generated by the tree grammar. Then, given regular tree languages L_1, L_2, ... L_n we propose an algorithm for generating a new type that corresponds to a tree language which contains the union of L_1, L_2, ... L_n *but which should be an LTG or a STTG*. Notice that even if the grammars $G_1, \ldots, G_n$ that generate $L_1, \ldots, L_n$ are LTGs (resp. STTGs), in general their union $G = G_1 \cup \cdots \cup G_n$ is not an LTG (resp. not a STTG) [Murata et al., 2005]. This is because the union of the sets of rules from these grammars may not respect the conditions imposed by the definitions of LTGs and STTGs.

In this context, our proposal can be formally expressed as follows: We present a method for extending a given regular tree grammar G into a new grammar G' respecting the following property: the language generated by G' is the smallest set of unranked trees that contains the language generated by G and the grammar G' is a Local Tree Grammar (LTG) or a Single-Type Tree Grammar (STTG).

Thus, the contribution of this paper is twofold:

1. We introduce two algorithms to transform a given regular grammar G into a new grammar G' such that: (i) $L(G) \subseteq L(G')$; (ii) G' is an LTG or a STTG; and (iii) $L(G')$ is the smallest language that respects constraints (*i*) and (*ii*). We offer formal proofs of some interesting properties of our methods.
2. We propose an interactive tool capable of guiding the user in the generation of a new type.

Paper organization: Section 2 gives an overview of our contributions. Section 3 introduces notations, concepts and properties needed in the paper. In Section 4

---

[2] http://relaxng.org/
[3] http://www.w3.org/TR/REC-xml/
[4] http://www.w3.org/XML/Schema

we consider the extension of a regular tree grammar to a local tree grammar while in Section 5 we deal with its extension to a single-type tree grammar. These proposals are revisited versions of the algorithms we have presented in [Chabin et al., 2010]. In Section 6, we modify the first algorithm to get an interactive tool for helping with the construction of XML types. In Section 7, we discuss time complexity and show some experiment results. Section 8 discusses some related work and Section 9 concludes the paper.

2 Overview

This section is an overview of our contributions. In our work, we consider the existence of an ontology associated to each grammar. An ontology alignment allows the implementation of translation functions that establish the correspondence among different words with the same meaning. In the following examples, we represent the tables (a kind of dictionary) over which the translation functions are implemented. All information in these tables is obtained from a given ontology alignment. A word without any entry in a table is associated to itself.

The first example illustrates the evolution of an XML type, expressed by a tree grammar, when the *resulting type should be a DTD (i.e., a LTG)*.

Example 1. Let G_1 be a regular tree grammar, resulting from the union of other regular tree grammars. We suppose that the data administrator needs a type for which both organisations of research laboratory are valid. He has an additional constraint: the resulting grammar should be expressed via an LTG (whose translation to a DTD is direct).

G_1		Translation table
$R_1 \rightarrow lab[T_1^*]$ $\qquad$ $R_2 \rightarrow lab[Emp^*]$		$researcher \leftrightarrow employee$
$T_1 \rightarrow team[Res^*]$ $\qquad$ $Emp \rightarrow employee[IsIn]$		$team \leftrightarrow group$
$Res \rightarrow researcher[\epsilon]$ $\qquad$ $IsIn \rightarrow group[\epsilon]$		

By translating the rules of G_1 according to the translation table we verify that the non-terminals Res and Emp generate the same terminal $researcher$, and consequently are in competition, which is forbidden for local tree grammars or DTDs (*i.e.*, G_1 is not an LTG). The same conclusion is obtained for non-terminals T_1 and $IsIn$ which generate terminal $team$. Non-terminals R_1 and R_2 are also competing ones. Our goal is to transform the new G_1 into an LTG, and this transformation will result in a grammar that generates a language L such that $L(G_1) \subseteq L$. In this context, we propose an algorithm that extends G_1 into a new local tree grammar G_A. The solution proposed is very simple. Firstly, rules concerning non-terminals Res and Emp are combined, given the rule $S \rightarrow researcher[IsIn \mid \epsilon]$. Secondly, in regular expressions (of other rules), Res and Emp should be replaced by S. Thus, we have $T_1 \rightarrow team[S^*]$ and $R_2 \rightarrow lab[S^*]$. All competing non-terminals are treated in the same way, giving rise to grammar G_A with rules: $R \rightarrow lab[T^* \mid S^*]$; $S \rightarrow researcher[T \mid \epsilon]$ and $T \rightarrow team[S^* \mid \epsilon]$. $\qquad\qquad$ □

The result obtained in Example 1 respects the imposed constraints: the obtained grammar is the least LTG (in the sense of language inclusion) and it generates a language that contains the language $L(G_1)$. This algorithm is presented in Section 4.

Next, we consider the situation where the resulting type is supposed to be specified by a XSD (or a single-type tree grammar). The following example illustrates an evolution in this context.

Example 2. Let us now consider G_2, a regular tree grammar resulting from the union of other regular tree grammars. We focus only on the rules concerning publication types. The translation table summarizes the correspondence among terminals. An ontology alignment associates *article* and *paper*.

G_2	Translation table
$R_3 \rightarrow article[Title.TitleJournal.Year.Vol]$	$article \leftrightarrow paper$
$R_4 \rightarrow paper[Title.TitleConf.Year]$	
$R_5 \rightarrow publication[(R_3 \mid R_4)^*]$	

We propose an algorithm that extends G_2 to a single type grammar G_B, which can then be translated into an XSD. Notice that the above rules of G_2 violate STTG constraints, since rule R_5 contains a regular expression with competing non-terminals. Indeed, rules $R_3 \rightarrow article[Title.TitleJournal.Year.Vol]$ and $R_4 \rightarrow paper[Title.TitleConf.Year]$ have equivalent terminals, according to the translation table. Thus, non-terminals R_3 and R_4 are competing ones. This is not a problem for a STTG. However, both non-terminals appear in the regular expression of rule R_5 and this is forbidden for a STTG. In this case, our method replaces the above rules by the following ones:

G_B
$R_6 \rightarrow paper[Title.TitleJournal.Year.Vol \mid Title.TitleConf.Year]$
$R_7 \rightarrow publication[R_6^*]$

$\square$

To get an LTG, competing non-terminals should be merged, which is simple. To get a STTG, the situation is more complicate: only competing non-terminals that appear in the same regular expression should be merged. Consequently, if R_1, R_2, R_3 are competing, but R_1, R_2 appear in the regular expression E whereas R_1, R_3 appear in E', then R_1 should be merged with R_2 (and not with R_3) within E, whereas R_1 should be merged with R_3 (and not with R_2) within E'. To do it in the general case, we introduce an equivalence relation over non-terminals, and consider equivalence classes as being the non-terminals of the new grammar. The algorithm is given in Section 5.

Now, suppose that the administrator wants to build a new type based on the types he knows, *i.e.*, by merging, in a flexible way, different types, *e.g.* G_3 and G_4. We propose a tool to guide him in this construction by considering one rule (of G_3 and G_4) at a time. The new type is built with the help of a dependency graph $D = (V, E)$. The set of nodes V contains the terminals of both grammars: the set of arcs E represents the notion of *dependency* among terminals in these grammars. The pair (a, b) of terminals is in E *iff* a production rule generating a contains a non-terminal which is associated to the terminal b. We just consider grammars where the dependency graph has no cycles. The following example illustrates this contribution.

Example 3. Let us consider the two grammars below. We concentrate in the rule concerning researchers.

G_3	G_4
$R_1 \rightarrow researcher[IdR.Name.Pub]$	$R_2 \rightarrow researcher[Name.Team.Ref]$

Figure 1 shows the dependency graph for these rules. We consider that non-terminals *IdR, Name, Pub, Name, Team, Ref* are the left-hand side of rules whose right-hand side contain terminals *idR, name, publications, team* and *references*, respectively. Arcs are coloured according to the grammar they come from: red (filled arrow) to indicate they come from both grammars, blue (dashed arrow) only from G_3 and green (dotted arrow) only from G_4.

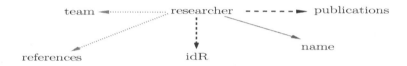

Fig. 1. Dependency graph for grammars G_3 and G_4

Our interactive tool proposes to follow this graph in a topological order: start with nodes with no output arcs, process them, delete them from the graph together with their input arcs, and so on. Processing a node here means writing its production rule. For each competing terminal, the user can propose a regular expression to define it. This regular expression is built only with non-terminals appearing as left-hand side of production rules already defined.

For instance, in our case, our interactive tool starts by proposing rules for each $A \in \{IdR, Name, Pub, Team, Ref\}$ (here, we consider that all these non-terminals are associated to rules of the format $A \rightarrow a[\epsilon]$). Then, the administrator can define the local type R, for researchers by using any of the above non-terminals. Suppose that the chosen rule is: $R \rightarrow researcher[(Name.Team.Pub)]$.

After this choice (and since in our example, no other type definition is expected), all rules defining useless types are discarded (*i.e.*, rules for *IdR* and *Ref* are discarded). Thus, we obtain a new grammar G_B, built by the administrator, guided by our application. More details of this tool are presented in Section 6.

Finally, if our administrator needs to accept documents of types G_3 or G_4 or G_B, he may use the first algorithm to define: $R \rightarrow researcher[(IdR.Name.Pub) \mid (Name.Team.Ref) \mid (Name.Team.Pub)]$.   □

The previous examples illustrate that our contribution is twofold. On the one hand, to propose algorithms that automatically compute a minimal type extension of given types. On the other hand to apply these algorithms as guides to allow the interactive definition of new types.

3   Theoretical Background

It is a well known fact that type definitions for XML and regular tree grammars are similar notions and that some schema definition languages can be represented by using specific classes of regular tree grammars. Thus, DTD and XML Schema, correspond, respectively, to Local Tree Grammars and Single-Type Tree Grammars [Murata et al., 2005]. Given an XML type T and its corresponding tree grammar G, the set of XML documents described by the type T corresponds to the language (set of trees) generated by G.

An XML document is an unranked tree, defined in the usual way as a mapping t from a set of positions (nonempty and closed under prefixes) $Pos(t)$ to an alphabet Σ. For $v \in Pos(t)$, $t(v)$ is the label of t at the position v, and $t|_v$ denotes the subtree of t at position v. Positions are sequences of integers in $\mathbb{N}^*$ and the set $Pos(t)$ satisfies: $j \geq 0, u.j \in Pos(t), 0 \leq i \leq j \Rightarrow u.i \in Pos(t)$, where the "." denotes the concatenation of sequences of integers. As usual, ϵ denotes the empty sequence of integers, *i.e.* the root position. The following figure shows a tree whose alphabet is the set of element names appearing in an XML document. In this case we have $t(\epsilon) = directory$, $t(0) = student$ and so on.

Given a tree t we denote by $t|_p$ the subtree whose root is at position $p \in Pos(t)$,*i.e.*, $Pos(t|_p) = \{s \mid p.s \in Pos(t)\}$ and for each $s \in Pos(t|_p)$ we have $t|_p(s) = t(p.s)$.

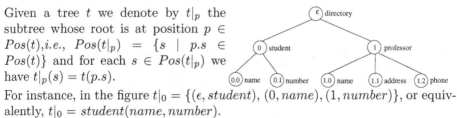

For instance, in the figure $t|_0 = \{(\epsilon, student), (0, name), (1, number)\}$, or equivalently, $t|_0 = student(name, number)$.

Given a tree t such that the position $p \in Pos(t)$ and a tree t', we note $t[p \leftarrow t']$ as the tree that results of substituting the subtree of t at position p by t'.

Definition 1 (Sub-tree, Forest). Let L be a set of trees. $ST(L)$ is the set of sub-trees of elements of L, i.e. $ST(L) = \{t \mid \exists u \in L, \exists p \in Pos(u), t = u|_p\}$. A *forest* is a (possibly empty) tuple of trees. For $a \in \Sigma$ and a forest $w = \langle t_1, \ldots, t_n \rangle$, $a(w)$ is the tree defined by $a(w) = a(t_1, \ldots, t_n)$. On the other hand, $w(\epsilon)$ is defined by $w(\epsilon) = \langle t_1(\epsilon), \ldots, t_n(\epsilon) \rangle$, *i.e.*, the tuple of the top symbols of w. □

Definition 2 (Regular Tree Grammar, Derivation). A *regular tree grammar* (RTG) is a 4-tuple $G = (N, T, S, P)$, where: N is a finite set of *non-terminal symbols*; T is a finite set of *terminal symbols*; S is a set of *start symbols*, where $S \subseteq N$ and P is a finite set of *production rules* of the form $X \to a[R]$, where $X \in N$, $a \in T$, and R is a regular expression over N (We say that, for a production rule, X is the left-hand side, $a[R]$ is the right-hand side, and R is the content model).

For an RTG $G = (N, T, S, P)$, we say that a tree t built on $N \cup T$ derives (in one step) into t' iff (i) there exists a position p of t such that $t|_p = A \in N$ and a production rule $A \to a\,[R]$ in P, and (ii) $t' = t[p \leftarrow a(w)]$ where $w \in L(R)$ ($L(R)$ is the set of words of non-terminals generated by R). We write $t \to_{[p, A \to a\,[R]]} t'$. More generally, a derivation (in several steps) is a (possibly empty) sequence of one-step derivations. We write $t \to_G^* t'$.

The *language* $L(G)$ generated by G is the set of trees containing only terminal symbols, defined by: $L(G) = \{t \mid \exists A \in S, A \to_G^* t\}$.   □

Remark. As usual, in this paper, our algorithms start from grammars in reduced form and (as in [Mani and Lee, 2002]) in normal form. A *regular tree grammar* (RTG) is said to be in **reduced form** if (i) every non-terminal is reachable from a start symbol, and (ii) every non-terminal generates at least one tree containing only terminal symbols. A regular tree grammar (RTG) is said to be in **normal form** if distinct production rules have distinct left-hand-sides.   □

To distinguish among sub-classes of regular tree grammars, we should understand the notion of competing non-terminals. Moreover, we define an equivalence relation on the non-terminals of grammar G, so that all competing non-terminals are in the same equivalence class. In our schema evolution algorithms (Sections 4 and 5), these equivalence classes form the non-terminals of the new grammar.

Definition 3 (Competing Non-terminals). Let $G = (N, T, S, P)$ be a regular tree grammar. Two non-terminals A and B are said to be *competing with each other* if $A \neq B$ and G contains production rules of the form $A \to a[E]$ and $B \to a[E']$ (i.e. A and B generate the same terminal symbol).

Define a **grouping relation over competing non-terminals** as follows. Let $\|$ be the relation on N defined by: for all $A, B \in N$, $A \parallel B$ iff $A = B$ or A and B are competing in P. For any $\chi \subseteq N$, let $\|_\chi$ be the restriction of $\|$ to the set χ ($\|_\chi$ is defined only for elements of χ).   □

Lemma 1. *Since G is in normal form, $\|$ is an equivalence relation. Similarly, $\|_\chi$ is an equivalence relation for any $\chi \subseteq N$.*   □

Definition 4 (Local Tree Grammar and Single-Type Tree Grammar). A *local tree grammar* (LTG) is a regular tree grammar that does not have competing non-terminals. A *local tree language* (LTL) is a language that can be generated by at least one LTG. A *single-type tree grammar* (STTG) is a regular tree grammar in normal form, where (i) for each production rule, non terminals in its regular expression do not compete with each other, and (ii) starts symbols do not compete with each other. A *single-type tree language* (STTL) is a language that can be generated by at least one STTG.   □

In [Murata et al., 2005] the expressive power of these classes of languages is discussed. We recall that LTL $\subset$ STTL $\subset$ RTL (*RTL for regular tree language*). Moreover, the LTL and STTL are closed under intersection but not under union; while the RTL are closed under union, intersection and difference. Note that converting an LTG into normal form produces an LTG as well.

4 Transforming an RTG into an LTG

Given a regular tree grammar $G_0 = (N_0, T_0, S_0, P_0)$, we propose a method to compute a local tree grammar G that generates the least local tree language containing $L(G_0)$.

In [Chabin et al., 2010], we have introduced a very intuitive version of our algorithm: replace each pair of competing non-terminals by a new non-terminal, until there are no more competing non-terminals.

In this section, we prefer to use the well-known formalism of equivalence classes (Lemma 1), which makes the proofs simpler, and allows an uniform notation $w.r.t.$ to the algorithm in the next section. Competing non-terminals are grouped together within an equivalence class, and the equivalence classes are the non-terminals of the new grammar G.

Algorithm 1 (RTG into LTG Transformation)
Notation:
(i) For any $A \in N_0$, $\hat{A}$ denotes the equivalence class of A w.r.t. relation $\|$, $i.e.$, $\hat{A}$ contains A and the non-terminals that are competing with A in P_0.
(ii) For any regular expression R, $\hat{R}$ is the regular expression obtained from R by replacing each non-terminal A by $\hat{A}$.
(iii) As usual, $N_0/_{\|}$ denotes the quotient set, i.e. the set of the equivalence classes.

Let $G_0 = (N_0, T_0, S_0, P_0)$ be a regular tree grammar. We define a new regular tree grammar $G = (N, T, S, P)$, obtained from G_0, as follows:
   Let $G = (N_0/_{\|}, T_0, S, P)$ where:

$-\ S = \{\hat{A} \mid A \in S_0\}$,
$-\ P = \{\ \{A_1, \ldots, A_n\} \rightarrow a\,[\hat{R}] \mid \{A_1, \ldots, A_n\} \in N_0/_{\|},$
$\qquad$ and $A_1 \rightarrow a[R_1], \ldots, A_n \rightarrow a[R_n] \in P_0$, and $R = (R_1 | \cdots | R_n)$. $\square$

The following example illustrates our algorithm.

Example 4. Let us consider merging the rules for two different DTDs for cooking recipes. Assuming that the vocabulary translations have already been done (on the basis of an alignment ontology), we build the grammar G_0 below. Each $A \in \{Name,\ Number,\ Unit,\ Quantity,\ Step,\ Item\}$ is associated to a production rule having the format $A \rightarrow a[\epsilon]$, meaning that label a is attached to data.

$$Recipe_a \rightarrow r[Ingreds.Recipe_a^*.Instrs_a] \qquad Ingreds \rightarrow is[OneIng_a^*]$$
$$OneIng_a \rightarrow ing[Name.Unit.Quantity] \qquad Instrs_a \rightarrow ins[Step^*]$$

$$Recipe_b \rightarrow r[Required.OneIng_b^*.Instrs_b] \qquad Required \rightarrow req[Item^*]$$
$$OneIng_b \rightarrow ing[Name.Quantity.Unit] \qquad Instrs_b \rightarrow ins[(Number.Step)^*]$$

Clearly, non-terminals $Recipe_a$ and $Recipe_b$, $OneIng_a$ and $OneIng_b$, $Instrs_a$ and $Instrs_b$ are competing. The equivalence classes for G_0 are $\{Recipe_a,\ Recipe_b\}$, $\{OneIng_a,\ OneIng_b\}$, $\{Instrs_a,\ Instrs_b\}$, $\{Ingreds\}$, $\{Required\}$, $\{Number\}$,

$\{Name\}, \{Unit\}, \{Quantity\}, \{Step\}, \{Item\}$. Each equivalence class is now seen as a new non-terminal. Our algorithm combines rules of G_0 whose left-hand non-terminals (in N_0) are in the same equivalence class. The obtained result is the LTG G below. To shorten the notations, for each non-terminals like X, Y_a, Y_b we write X instead of $\{X\}$, and Y instead of $\{Y_a, Y_b\}$. The missing rules are of the form $A \to a[\epsilon]$.

$$Recipe \to r[(Ingreds.Recipe^*.Instrs)|(Required.OneIng^*.Instrs])$$
$$Ingreds \to is[OneIng^*]$$
$$OneIng \to ing[(Name.Unit.Quantity)|(Name.Quantity.Unit)]$$
$$Instrs \to ins[Step^*|(Number.Step)^*]$$
$$Required \to req[Item^*]$$

$\square$

One of the most important features of our algorithm is its simplicity. However, one fundamental contribution of this paper is the proof that, with this very simple method, we can compute the smallest extension of a given tree grammar, respecting the constraints imposed on an LTG. This result is stated in the following theorem.

Theorem 1. *The grammar returned by Algorithm 1 generates the least LTL that contains $L(G_0)$.* $\square$

The intuition behind the proof of Theorem 1 is as follows. Let G be the grammar produced by our algorithm and let G' be any LTG such that $L(G_0) \subseteq L(G')$, we have to prove that $L(G_0) \subseteq L(G)$ (soundness), and that $L(G) \subseteq L(G')$ (minimality: $L(G)$ is the least LTL containing $L(G_0)$). Proving soundness is not very difficult. Minimality comes from the following steps: **(A)** As production rules of an LTG in normal form define a bijection between the sets of terminals and non-terminals, there is only one rule in G of the form $\hat{A}_1 \to a[R]$ producing subtrees with root a. By the construction of our algorithm this rule should correspond to rules $A_i \to a[R_i]$ in G_0 with $i \in \{1, \cdots, n\}$. All A_i are competing in G_0 and no other symbol in N_0 is competing with a A_i so $\hat{A}_1 = \cdots = \hat{A}_n = \{A_1, \cdots, A_n\}$. And we have $R = \hat{R}_1 | \cdots | \hat{R}_n$. **(B)** Consequently, we can prove that if $a(w)$ is a subtree of $t \in L(G)$, then there is at least one tree in $L(G_0)$ with $a(w')$ as a subtree, s.t. $w'(\epsilon) = w(\epsilon)$ (i.e. forests w' and w have the same tuple of top-symbols). **(C)** w is a forest composed by subtrees of $L(G)$, and by induction hypothesis applied to each component of w (each component is a strict subtree of $a(w)$), we know that w is also a forest composed by subtrees of $L(G')$. On the other hand, since $L(G_0) \subseteq L(G')$, $a(w')$ is a subtree of $L(G')$. **(D)** As G' is an LTG, and thanks to some properties of local languages, we can replace each subtree of $a(w')$, rooted by the elements of $w'(\epsilon)$, by the corresponding subtree of $a(w)$ and thus, $a(w)$ is a subtree of $L(G')$. **(E)** Finally, as this is valid for every subtree, we have that $L(G) \subseteq L(G')$.

Appendix A presents the proof of Theorem 1.

5   Transforming an RTG into a STTG

Given a regular tree grammar $G_0 = (N_0, T_0, S_0, P_0)$, the following algorithm computes a single-type tree grammar G that generates the least single-type tree language containing $L(G_0)$. It is based on grouping competing non-terminals into equivalence classes, in a way different from Algorithm 1. Here, we group competing non-terminals $A_1, \ldots, A_n$ together, only if they appear in the same regular expression R of G_0, and in this case the set $\{A_1, \ldots, A_n\}$ is a non-terminal of G. If $A_1, \ldots, A_n$ do not appear in the same regular expression, we have to consider subsets of $\{A_1, \ldots, A_n\}$. This is why Algorithm 2 (and proofs) is more complicated than Algorithm 1.

Algorithm 2 (RTG into STTG Transformation)
Notation:
(i) For any regular expression R, $N(R)$ denotes the set of non-terminals occurring in R.
(ii) For any $\chi \subseteq N_0$ and any $A \in \chi$, $\hat{A}^\chi$ denotes the equivalence class of A w.r.t. relation $\|_\chi$, i.e. $\hat{A}^\chi$ contains A and the non-terminals of χ that are competing with A in P_0.
(iii) $\sigma_{N(R)}$ is the substitution defined over $N(R)$ by $\forall A \in N(R)$, $\sigma_{N(R)}(A) = \hat{A}^{N(R)}$. By extension, $\sigma_{N(R)}(R)$ is the regular expression obtained from R by replacing each non-terminal A in R by $\sigma_{N(R)}(A)$.

Let $G_0 = (N_0, T_0, S_0, P_0)$ be a regular tree grammar. We define a new regular tree grammar $G = (N, T, S, P)$, obtained from G_0, according to the following steps:

1. Let $G = (\mathcal{P}(N_0), T_0, S, P)$ where:
   - $S = \{\hat{A}^{S_0} \mid A \in S_0\}$,
   - $P = \{\, \{A_1, \ldots, A_n\} \to a\,[\sigma_{N(R)}(R)] \mid$
     $A_1 \to a[R_1], \ldots, A_n \to a[R_n] \in P_0, R = (R_1 | \cdots | R_n)\}$,
     where $\{A_1, \ldots, A_n\}$ indicates all non-empty sets containing competing non-terminals (not only the maximal ones).
2. Remove unreachable non-terminals and unreachable rules in G; return G. $\square$

The difference between STTG and LTG versions (Section 4) is in the use of non-maximal sets of competing non-terminals. In particular, Algorithm 2 considers (step 1) *each* set of competing non-terminals as a left-hand side (and not only maximal sets) to build the production rules of G. Thus, at step 1, G may create unreachable rules (from the start symbols), which are then removed at step 2. Algorithm 2 eases our proofs. An optimized version, where just the needed non-terminals are generated, is given in [Chabin et al., 2010].

The following example illustrates that for an STTG only competing non-terminals appearing in the same regular expression are combined to form a new non-terminal.

Example 5. Let G_0 be a non-STTG grammar having the following set P_0 of productions rules (*School* is the start symbol). It describes a French school with students enrolled to an international English section (*IntStudent*) and normal French students (*Student*). Different options are available for each student class.

$$School \to school[IntStudent \mid Student]$$
$$Student \to student[Name.Option3]$$
$$IntStudent \to intstudent[Name.(Option1 \mid Option2)]$$

$$Option1 \to option[EL.GL] \qquad\qquad Option2 \to option[EL.SL]$$
$$Option3 \to option[EL] \qquad\qquad Name \to name[\epsilon]$$
$$EL \to english[\epsilon] \qquad\qquad GL \to german[\epsilon]$$
$$SL \to spanish[\epsilon]$$

The grammar G obtained by our approach has the rules below where non terminals are named by their equivalence class. Clearly, they can be denoted by shorter notations.

$$\{School\} \to school[\{IntStudent\} \mid \{Student\}]$$
$$\{IntStudent\} \to intstudent[\{Name\}.\{Option1, Option2\}]$$
$$\{Student\} \to student[\{Name\}.\{Option3\}]$$
$$\{Option1, Option2\} \to option[(\{EL\}.\{GL\}) \mid (\{EL\}.\{SL\})]$$
$$\{Option3\} \to option[\{EL\}]$$
$$\{Name\} \to name[\epsilon]$$
$$\{EL\} \to english[\epsilon]$$
$$\{GL\} \to german[\epsilon]$$
$$\{SL\} \to spanish[\epsilon]$$

Notice that although *Option1*, *Option2* and *Option3* are competing non-terminals; our approach does not produce new non-terminals corresponding to the combination of all of them. For instance, *Option1* and *Option2* are combined in order to generate the non-terminal $\{Option1, Option2\}$, but we do not need to produce a non-terminal $\{Option1, Option3\}$ since *Option1* and *Option3* do not appear together in a regular expression. We also generate $\{Option3\}$ as non-terminal because *Option3* is alone in the rule defining *Student*.   □

The following example offers an interesting illustration of the extension of the original language.

Example 6. Consider a non-STTG grammar G_0 having the following set P_0 of productions rules (*Image* is the start symbol):

$$Image \to image[Frame1 \mid Frame2 \mid Background.Foreground]$$
$$Frame1 \to frame[Frame1.Frame1 \mid \epsilon]$$
$$Frame2 \to frame[Frame2.Frame2.Frame2 \mid \epsilon]$$
$$Background \to back[Frame1]$$
$$Foreground \to fore[Frame2]$$

Grammar G_0 defines different ways of decomposing an image: recursively into two or three frames or by describing the background and the foreground separately. Moreover, the background (resp. the foreground) is described by binary decompositions (resp. ternary decompositions). In other words, the language of G_0 contains the union of the trees: *image(bin(frame))*; *image(ter(frame))* and *image (back (bin (frame)), fore (ter (frame)))* where *bin* (resp. *ter*) denotes the set of all binary (resp. ternary) trees that contains only the symbol *frame*. The result is G, which contains the rules below (the start symbol is $\{Image\}$):

$$\{Image\} \rightarrow image[\{Frame1, Frame2\} \mid \{Background\}.\{Foreground\}]$$
$$\{Background\} \rightarrow back[\{Frame1\}]$$
$$\{Foreground\} \rightarrow fore[\{Frame2\}]$$
$$\{Frame1, Frame2\} \rightarrow frame[\epsilon \mid \{Frame1, Frame2\}.\{Frame1, Frame2\}$$
$$\mid \{Frame1, Frame2\}.\{Frame1, Frame2\}.\{Frame1, Frame2\}]$$
$$\{Frame1\} \rightarrow frame[\{Frame1\}.\{Frame1\} \mid \epsilon]$$
$$\{Frame2\} \rightarrow frame[\{Frame2\}.\{Frame2\}.\{Frame2\} \mid \epsilon]$$

Grammar G is a STTG that generates the union of *image(tree(frame))* and *image (back (bin (frame)), fore (ter (frame)))* where *tree* denotes the set of all trees that contain only the symbol *frame* and such that each node has 0 or 2 or 3 children. Let $L_G(X)$ be the language obtained by deriving in G the non-terminal X. Actually, $L_G(\{Frame1, Frame2\})$ is the least STTL that contains $L_{G_0}(Frame1) \cup L_{G_0}(Frame2)$. □

An important part of our work consist in proving the following theorem. We have presented part of this proof in [Chabin et al., 2010], but the interested reader can find its complete version in Appendix B.

Theorem 2. *The grammar returned by Algorithm 2 generates the least STTL that contains $L(G_0)$.* □

From this result, we are able to ensure that our algorithm generates the least STTL responding to our goals. This is an important result when dealing with type we want to extend in a controlled way.

6   Interactive Generation of New Types

This section introduces an interactive tool which may help an advised user to build an XML type based on existing ones. This tool is useful when an administrator decides to propose a new type which should co-exist with other (similar) types for a certain time. For instance, the description of research laboratories in a university may vary. Our administrator wants to organize information in a more uniform way by proposing a new type, a schema evolution, based on the existing types (since old types represent all information available and needed). In this paper we outline the main ideas of our tool which can be used in different contexts where slightly different XML types exist and should be catalogued. Indeed, this kind of application needs the two parts of our proposal: to extend the union of types to a standard XML schema language and to interactively allow the construction of a new type. We illustrate the interactive function of our tool (outlined in Section 1) in a more complete example.

Example 7 (Interactive approach). We consider the rules from two LTG after translating terminals into a unified nomenclature, according to a translation table. Each grammar shows a different organization of research laboratories. In G_1 laboratories are organized with researchers and teams. Publications are sub-elements of researchers. Teams are composed by researchers, identified by their identification numbers. All $A \in \{Dom, IdR, First, Last, TitleP, TitleConf, Year, Vol\}$ are associated to production rules having the format $A \rightarrow a[\epsilon]$.

$Lab \rightarrow lab[Dom^*.R^*.Team^*]$ $R \rightarrow researcher[IdR.Name.P]$
$P \rightarrow publications[CPaper^*.JPaper^*]$ $Team \rightarrow team[IdR^*]$
$CPaper \rightarrow confPaper[TitleP.TitleConf.Year]$
$JPaper \rightarrow jourPaper[TitleP.TitleJ.Year.Vol]$
$Name \rightarrow name[First.Last]$

Grammar G_2 represents an organization where researchers and publications are stored in an independent way, but where references are supposed to link informations. A team is just a sub-element of a researcher. All $A \in \{Dom, Name, Code, TitleP, TitleConf, Year, Vol\}$ are associated to production rules having the format $A \rightarrow a[\epsilon]$.

$Lab \rightarrow lab[Dom.R^*.P]$ $R \rightarrow researcher[Name.Team.Ref]$
$Ref \rightarrow references[Code^*]$ $P \rightarrow publications[CPaper^*.JPaper^*]$
$ConfPaper \rightarrow confPaper[Code.TitleP.TitleConf.Year]$
$JPaper \rightarrow jourPaper[Code.TitleP.TitleJ.Year.Vol]$

To guide the construction of a new grammar we use a dependency graph. Figure 2 shows the dependency graph for grammars G_1 and G_2.

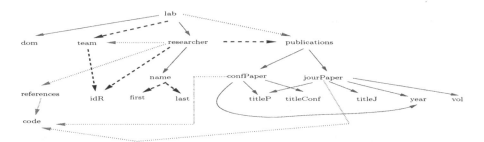

Fig. 2. Dependency graph for grammars G_1 and G_2

As mentioned in Example 3, to aid an advised user in the construction of a new XML type, we propose to follow this graph in a topological order and, for each competing terminal, to ask the user a regular expression to define it. This regular expression is built only by non-terminals already defined. From Figure 2 the user starts by considering terminals $a \in \{idR, name, team, titleP, titleConf, title.J, year, code, vol\}$. Let G_3 be the grammar obtained by this interactive method. Clearly, all terminals a are associated to *data* and thus the corresponding grammar rules are of the format $A \to a[\epsilon]$. The following production rules correspond to the user choices.

$Lab \to lab[Dom.R^*]$ $R \to researcher[Name.Team.P]$
$P \to publications[CPaper^*.JPaper^*]$
$CPaper \to confPaper[TitleP.TitleConf.Year]$
$JPaper \to jourPaper[TitleP.TitleJ.Year.Vol]$ □

We now formally present our interactive algorithm (which adapts Algorithm 1 to an interactive context). We recall that for a regular expression R (resp. a tree t), $NT(R)$ denotes the set of non-terminals occurring in R (resp. in t).

Definition 5 (Order Relation over Terminals). Let $G = (N, T, S, P)$ be an RTG. Let $a, b \in T$. We define the relation $\leadsto_G$ over terminals by $a \leadsto_G b$ iff there exist production rules $A \to a[R]$, $B \to b[R']$ in P such that $B \in NT(R)$. In other words, b may be a child of a.   □

Non-recursivity is defined as usual, i.e. over non-terminals: it means that a non-terminal A can never derive a tree that contains A again. Using terminals, we get a stronger property, called strong non-recursivity, which means that the dependency graph is acyclic.

Definition 6 (Recursivity)
A grammar G is *non-recursive* iff $\neg(\exists A \in N, A \to_G^+ t \wedge A \in NT(t))$, where t is a tree. A grammar G is *strongly non-recursive* iff $\neg(\exists a \in T, a \leadsto_G^+ a)$, where $\to_G^+$ and $\leadsto_G^+$ are the transitive closures of $\to_G$ and $\leadsto_G$, respectively.   □

Lemma 2. *If G is strongly non-recursive, then G is non-recursive.*
Proof: By contraposition. Suppose $\exists A_0 \in N, A_0 \to_G^+ t \wedge A_0 \in NT(t))$. Then: $\exists n \in \mathbb{N}\backslash\{0\}, \forall i \in \{0, \ldots, n-1\}, \exists A_i \to a_i[R_i] \in P, A_{i+1} \in NT(R_i) \wedge A_n = A_0$. By definition of $\leadsto_G$ we have: $a_0 \leadsto_G a_1 \leadsto_G \cdots \leadsto_G a_{n-1} \leadsto_G a_0$.   □

Lemma 3. *If G is strongly non-recursive, then $\leadsto_G^+$ is a strict (partial) order.*
Proof: $\leadsto_G^+$ is transitive. On the other hand, for all $a, b \in T, a \leadsto_G^+ b \wedge b \leadsto_G^+ a$ implies $a \leadsto_G^+ a$, which is impossible since G is strongly non-recursive.   □

Algorithm 3 (Interactive Generation of an LTG)
Notation: Let $G_0 = (N_0, T_0, S_0, P_0)$ be a regular tree grammar[5] strongly non-recursive. For each terminal a, consider all the rules in P_0 that generates a, say $A_1 \to a[R_1], \ldots, A_n \to a[R_n]$. Then we define $\hat{a} = \{A_1, \ldots, A_n\}$. Note that $\hat{a} \in N_0/_{\parallel}$, i.e. $\hat{a}$ is an equivalence class.

We define a new regular tree grammar $G = (N, T, S, P)$, obtained from G_0, according to the steps:

---

[5] Recall that G_0 is in reduced normal form and thus, for each $A \in N_0$ there exists exactly one rule in P_0 whose left-hand-side is A.

1. Let $G = (N_0/_{\|}, T_0, S, P)$ where:
   - $S = \{\hat{A} \mid A \in S_0\}$,
   - $P = \{\hat{a} \rightarrow a\,[R] \mid a \in T_0,$

     and $\hat{a} = \{A_1, \ldots, A_n\} \in N_0/_{\|}$,

     and $A_1 \rightarrow a[R_1], \ldots, A_n \rightarrow a[R_n] \in P_0$,

     and (i) $R = (\hat{R}_1 | \cdots | \hat{R}_n)$ or (ii) R is defined by the user s.t.

     $$\forall B \in NT(R),\ B = \hat{b} \wedge a \rightsquigarrow^+_{G_0} b.$$

2. Remove all unreachable or unproductive non-terminals and rules in G, then return it.   □

The aiding tool for XML type construction we propose is based on Algorithm 3. However, to make it more user friendly, each time a user wants to propose a new local type (the interactive step mentioned in item 1(ii)), some facilities are offered. The first one aims at releasing the user of thinking about grammar "technical problems" and distinction concerning terminal and non-terminals. Therefore, our tool allows the user to propose regular expressions built over XML labels (*i.e.*, the terminals of G). Indeed, this opportunity matches the use of DTDs. Grammar G resulting from Algorithm 3 is automatically obtained by replacing each terminal b (used by the user in the construction of the type) by non-terminal $\hat{b}$. Note that the limitation of the user choice in item 1(ii) (only b's s.t. $a \rightsquigarrow^+_{G_0} b$ are allowed) is necessary to prevent from introducing cycles in the dependency graph, i.e. to get a strongly non-recursive grammar. The second facility aims at guiding the user in a good definition order. Thus, at each step, our tool may guide the user to choose new (local) types according to the order established by a topological sort of the dependency graph: one may choose the type of a terminal a once the type of every $b \in T_0$ such that $a \rightsquigarrow^+_{G_0} b$ has already been treated (bottom-up approach).

We are currently discussing some other improvements to our tool. As a short term optimisation, we intend to allow a global design of an XML type before using Algorithm 3. By using a graphical interface, the user can, in fact, transform the dependency graph into a tree. In this way, he establishes a choice before entering in the details of each local type. For instance, in Example 7, Figure 2, terminal *team* has two parents, meaning that it can take part in the definition of *researcher* or *laboratory*. However, a user probably wants to choose one of these possibilities and not use *team* in both definitions (which is allowed by our algorithm), to avoid redundancy. By deleting the arc between *lab* and *team*, the user fixes, beforehand, his choice, avoiding useless computation. We currently consider the existence of an ontology alignment from which we can obtain a translation table for different terminals used in the grammars. A long term improvement concerns the methods to automatically generate this table. We can fin in [Gu et al., 2008] some initial clues to deal with this aspect.

Now we prove that grammars obtained by Algorithm 3 are strongly non-recursive LTGs.

Lemma 4. $\forall a, b \in T_0,\ a \rightsquigarrow_G b \implies a \rightsquigarrow^+_{G_0} b.$ □

Proof: Suppose $a \leadsto_G b$. Then there exist rules $\hat{a} \to a[R]$, $\hat{b} \to b[R'] \in P$ s.t. $\hat{b} \in NT(R)$. Therefore there exist $A_1 \to a[R_1], \ldots, A_n \to a[R_n] \in P_0$ and $B_1 \to b[R'_1], \ldots, B_k \to b[R'_k] \in P_0$ s.t. $\hat{a} = \{A_1, \ldots, A_n\}$ and $\hat{b} = \{B_1, \ldots, B_k\}$. To build rule $\hat{a} \to a[R]$, there are two possibilities:

Case (i): $R = (\hat{R}_1 | \cdots | \hat{R}_n)$. Since $\hat{b} \in NT(R)$, there exists $j \in \{1, \ldots, n\}$ s.t. $\hat{b} \in NT(\hat{R}_j)$. Then $\exists p \in \{1, \ldots, k\}$, $B_p \in NT(R_j)$. Finally we have $A_j \to a[R_j] \in P_0$, $B_p \to b[R'_p] \in P_0$, and $B_p \in NT(R_j)$. Consequently $a \leadsto_{G_0} b$.

Case (ii) $\forall C \in NT(R)$, $C = \hat{c} \wedge a \leadsto^+_{G_0} c$. Since $\hat{b} \in NT(R)$, we have $a \leadsto^+_{G_0} b$. □

Theorem 3. *The grammar returned by Algorithm 3 is a strongly non-recursive LTG in normal form.*   □

Proof: In Algorithm 3, for each $a \in T_0$ we define only one rule in P that generates a: it is the rule $\hat{a} \to a[R]$. Therefore there are no competing non-terminal in G, then G is an LTG. On the other hand, suppose that G is not in normal form, i.e. $\exists b \in T_0$, $b \neq a \wedge \hat{b} \to b[R'] \in P \wedge \hat{b} = \hat{a} = \{C_1, \ldots, C_n\}$. Then for all $i \in \{1, \ldots, n\}$, there exist two rules $C_i \to a[R_i]$, $C_i \to b[R'_i] \in P_0$, which is impossible since G_0 is in normal form.

Suppose that G is not strongly non-recursive. Then $\exists a \in T_0$, $a \leadsto^+_G a$. From Lemma 4, we get $a \leadsto^+_{G_0} a$ which is impossible since G_0 is strongly non-recursive.   □

7   Algorithm Analysis and Experiments

Algorithm 1 is polynomial. Recall that, for each original rule, the algorithm verifies whether there exist competing non-terminals by visiting the remaining rules and merges rules where competition is detected. The algorithm proceeds by traversing each regular expression of the rules obtained from this first step, replacing each non-terminal by the associated equivalent class. Thus, in the worst case, Algorithm 1 runs in time $O(N^2 + N.l)$, where N is the number of production rules and l is the maximal number of non-terminals in a regular expression.

In the following we consider an example which illustrates the worst case for Algorithm 2. Indeed, in the worst case, the number of non-terminals of the STTG returned by Algorithm 2 is exponential in the number of the non-terminals of the initial grammar G_0. However, in real cases, it is difficult to find such a situation.

Example 8. Let us consider a grammar where the production rules have the following form:

1	$S \to s[A_1|B_1|C_1]$		2	$A_1 \to a[A_2 \mid \epsilon]$
3	$A_2 \to a[A_1]$		4	$B_1 \to a[B_2 \mid \epsilon]$
5	$B_2 \to a[B_3]$		6	$B_3 \to a[B_1]$
7	$C_1 \to a[C_2 \mid \epsilon]$		8	$C_2 \to a[C_3]$
9	$C_3 \to a[C_4]$		10	$C_4 \to a[C_1]$

Clearly, this grammar is not a STTG, since in the first rule we have a regular expression with three competing non-terminals. By using Algorithm 2, the first rule is transformed into $\{S\} \to s[\{A_1, B_1, C_1\} \mid \{A_1, B_1, C_1\} \mid \{A_1, B_1, C_1\}]$. Trying to merge rules 2, 5 and 7 we find $\{A_1, B_1, C_1\} \to a[\sigma(A_2|B_2|C_2)]$ where the regular expression $A_2|B_2|C_2$ has also three competing non-terminals that should be *put together* to give a new non-terminal $\{A_2, B_2, C_2\}$. The reasoning goes on in the same way to obtain $\{A_1, B_3, C_3\}$, $\{A_2, B_1, C_4\}$ and so on. The number of non-terminals grows exponentially.   □

A prototype tool implementing Algorithm 1 and Algorithm 2 may be downloaded from [Chabin et al.,]. It was developed using the ASF+SDF Meta-Environment [van den Brand et al., 2001] and demonstrates the feasibility of our approach. To study the scalability of our method, Algorithm 1 has been implemented in Java. In this context, we have also developed some tools for dealing with tree grammars. Thus, given a tree grammar G, we can test whether it is in reduced form or in normal form or whether the grammar is already an LTG.

Table 1 shows the results of some experiments obtained by the Java implementation of Algorithm 1. Our experiments were done on an Intel Dual Core P8700, 2.53GHz with 2GB of memory. To perform these tests we have developed an RTG generator[6]: from n_1 terminals, we generate n_2 rules (or non-terminals, since each non-terminal is associated to one rule). When $n_1 \leq n_2$ the generated RTG has n_1 non-terminals that are not competing and $n_2 - n_1$ competing non-terminals. When $n_1 = n_2$ the generated grammar is an LTG. The regular expression of each rule has the form $E_1 \mid \cdots \mid E_n$ where each E_i is a conjunction of k non-terminals randomly generated and randomly adorned, or not, by $*$ or $?$. The values of n and k are also chosen at random, limited by given thresholds.

Table 1. Runtime for Algorithm 1 in milliseconds

Example number	Number of terminals	Number of non-terminals	Runtime for Algorithm 1 (ms) LTG transformation
1	250	300	349
2	250	500	536
3	250	1000	2316
4	1000	1000	1956
5	1000	2000	8522
6	2000	2000	8093
7	1000	4000	36349
8	1000	8000	163236
9	1000	10000	265414

From Table 1 it is possible to see that time execution increases polynomially according to the number of non-terminals (roughly, when the number of non-terminals is multiplied by 2, time is multiplied by 4). We can also notice that

---

[6] Available at `http://www.univ-orleans.fr/lifo/Members/chabin/logiciels.html`

when changing only the number of terminals, the impact of competition on execution time is not very important. For instance, lines 2 and 3 show grammars having the same set of terminals but a different number of non-terminals (or production rules). Clearly, grammar on line 3 needs much more time to be proceeded. However, lines 3 and 4 (or lines 5 and 6) show two grammars having the same set of non-terminals but a different number of terminals. In our examples, this fact indicates that grammar of line 3 has more competing non-terminals than grammar on line 4. Notice however both cases are treated in approximately the same time.

We have just developed prototypes over which we can evaluate the initial interest of our proposal. Even if some software engineer effort is necessary to transform these prototypes into a software tool, the performed tests on our Java implementation show that we can expect a good performance of our method in the construction of tools for manipulating and integrating XML types.

8   Related Work

XML type evolution receives more and more attention nowadays, and questions such as incremental re-validation ([Guerrini et al., 2005]), document correction w.r.t type evolution ([Amavi et al., 2013]) or the impact of the evolution on queries ([Genevès et al., 2009, Moro et al., 2007]) are some of the studied aspects. However, data administrators still need simple tools for aiding and guiding them in the evolution and construction of XML types, particularly when information integration is needed. This paper aims to respond to this demand.

Algorithms 1 and 2 allow the conservative evolution of schemas. Our work complements the proposals in [Bouchou et al., 2009, da Luz et al., 2007], since we consider not only DTD but also XSD, and adopts a global approach where all the tree grammar is taken into account as a whole. Our algorithms are inspired in some grammar inference methods (such as those dealing with ranked tree languages in [Besombes and Marion, 2003, Besombes and Marion, 2006]) that return a tree grammar or a tree automaton from a set of positive examples (see [Angluin, 1992, Sakakibara, 1997] for surveys). Our method deals with unranked trees, starts from a given RTG G_0 (representing a set of positive examples) and finds the least LTG or STTG that contains $L(G_0)$. As we consider an initial tree grammar we are not exactly inserted in the learning domain, but their methods inspire us and give us tools to solve our problem, namely, the evolution of a original schema (and not the extraction of a new schema).

In [Garofalakis et al., 2000, Bex et al., 2006, Bex et al., 2007] we find examples of work on XML schema inference . In [Bex et al., 2006] DTD inference consists in an inference of regular expressions from positive examples. As the seminal result from Gold [Gold, 1967] shows that the class of all regular expressions cannot be learnt from positive examples, [Bex et al., 2006] identifies classes of regular expressions that can be efficiently learnt. Their method is extended to deal with XMLSchema (XSD) in [Bex et al., 2007].

The approach in [Abiteboul et al., 2009] can be seen as the inverse of ours. Let us suppose a library consortium example. Their approach focus on defining the subtypes corresponding to each library supposing that a target global type of a distributed XML document is given. Our approach proposes to find the integration of different library subtypes by finding the least library type capable of verifying all library subtypes.

The usability of our method is twofold: as a theoretical tool, it can help answering the decision problem announced in [Martens et al., 2006]; as an applied tool, it can easily be adapted to the context of digital libraries, web services, etc. In [Martens et al., 2006], the authors are interested in analysing the actual expressive power of XSD. With some non-trivial amount of work, part of their theorem proofs can be used to produce an algorithm similar to ours. Indeed, in [Gelade et al., 2010] (a work simultaneous to ours in [Chabin et al., 2010]), the authors decide to revisit their results in [Martens et al., 2006] to define approximations of the union (intersection and complement) of XSD schemas. Our methods are similar, but our proposal works directly over grammars, allowing the implementation of a user friendly tool easily extended to an interactive mode, while results in [Gelade et al., 2010] are based on the construction of type automata.

A large amount of work have been done on the subject of matching XML schemas or ontology alignment ([Shvaiko and Euzenat, 2005] as a survey) and we can find a certain number of automatic tools for generating schema matchings such as SAMBO [Lambrix et al., 2008] or COMA++ [Maßmann et al., 2011]. Generally, a schema matching gives a set of edges, or correspondences, between pairs of elements, that can be stored into translation tables (a kind of dictionary). An important perspective of our work concerns the generation of translation tables by using methods such as the one proposed in [Gu et al., 2008], since, until now these semantics aspects have been considered as a 'given information'.

An increasing demand on data exchange and on constraint validation have motivated us to work on the generation of a new set of constraints from different local sets of type or integrity restrictions. This new set of constraints should keep all non contradictory local restrictions. The type evolution proposed here is well adapted to our proposes and it seems possible to combine it with an XFD filter, as the one in [Amavi and Halfeld Ferrari, 2012], in order to obtain a (general) set of constraints allowing interoperability. This paper focus only on schema constraints and proposes an extension that guarantees the validity of any local document. Thus, as explained in the introduction, our approach is very interesting when local systems $I_1, \ldots, I_n$, inter-operate with a global system I which should receive information from any local source (or format) and also ensure type constraint validation.

9   Conclusion

XML data and types age or need to be adapted to evolving environments. Different type evolution methods propose to trigger document updates in order to assure document validity. Conservative type evolution is an easy-to-handle evolution method that guarantees validity after a type modification.

This paper proposes conservative evolution algorithms that compute a local or single-type grammar which extends minimally a given original regular grammar. The paper proves the correctness and the minimality of the generated grammars. An interactive approach for aiding in the construction of new schemas is also introduced. Our three algorithms represent the basis for the construction of a platform whose goal is to support administration needs in terms of maintenance, evolution and integration of XML types. One possible application of our work is in the field of Digital Libraries, due to their need of evolution when new sources of data become available or when merging two libraries may be interesting [Crane, 2006]. In these cases, it is necessary to build a new schema for the data being merged.

We are currently working on improving and extending our approach to solve other questions related to type compatibility and evolution. Except for the terminal translation table, our approach is inherently syntactic: only structural aspects of XML documents are considered and our new grammars are built by syntactic manipulation of the original production rules. However, schemas can be more expressive than DTD and XSD, associated to integrity constraints (as in [Bouchou et al., 2012]) or expressed by a semantically richer data model (as in [Wu et al., 2001]).

In [Amavi and Halfeld Ferrari, 2012] we find an algorithm that computes, from given local sets of XFD, a cover of the biggest set of XFD that does not violate any local document. This algorithm is a first step towards an extension of our approach which will take into account integrity constraints. Notice that we understand this extension by the implementation of different procedures, one for each kind of integrity constraints. In other words, by using the uniform formalism proposed in [Bouchou et al., 2012] for expressing integrity constraints on XML documents and following the ideas exposed in [Amavi and Halfeld Ferrari, 2012], we can build sets of integrity constraints (inclusion dependencies, keys, etc) adapted to our global schema. In this way, the evolution of richer schema would correspond to the parallel evolution of different sets of constraints.

We intend not only to extend our work in these directions but also to enrich our platform with tools (such as the one proposed in [Amavi et al., 2011]) for comparing or classifying types with respect to a 'type distance' capable of choosing the closest type for a given document (as discussed, for instance, in [Tekli et al., 2011, Bertino et al., 2008]). Interesting theoretical and practical problems are related to all these perspectives.

References

[Abiteboul et al., 2009] Abiteboul, S., Gottlob, G., Manna, M.: Distributed xml design. In: PODS 2009: Proceedings of the Twenty-Eighth ACM SIGMOD-SIGACT-SIGART Symposium on Principles of Database Systems, pp. 247–258. ACM (2009)

[Amavi et al., 2013] Amavi, J., Bouchou, B., Savary, A.: On correcting XML documents with respect to a schema. The Computer Journal 56(4) (2013)

[Amavi et al., 2011] Amavi, J., Chabin, J., Halfeld Ferrari, M., Réty, P.: Weak Inclusion for XML Types. In: Bouchou-Markhoff, B., Caron, P., Champarnaud, J.-M., Maurel, D. (eds.) CIAA 2011. LNCS, vol. 6807, pp. 30–41. Springer, Heidelberg (2011)

[Amavi and Halfeld Ferrari, 2012] Amavi, J., Halfeld Ferrari, M.: An axiom system for XML and an algorithm for filtering XFD (also a poster published in sac 2013, Technical Report RR-2012-03, LIFO/Université d'Orléans (2012)

[Angluin, 1992] Angluin, D.: Computational learning theory: survey and selected bibliography. In: STOC 1992: Proceedings of the Twenty-Fourth Annual ACM Symposium on Theory of Computing, pp. 351–369. ACM, New York (1992)

[Bertino et al., 2008] Bertino, E., Giovanna Guerrini, G., Mesiti, M.: Measuring the structural similarity among XML documents and dtds. J. Intell. Inf. Syst. 30, 55–92 (2008)

[Besombes and Marion, 2003] Besombes, J., Marion, J.-Y.: Apprentissage des langages réguliers d'arbres et applications. Traitement Automatique de Langues 44(1), 121–153 (2003)

[Besombes and Marion, 2006] Besombes, J., Marion, J.-Y.: Learning tree languages from positive examples and membership queries. Theoretical Computer Science (2006)

[Bex et al., 2006] Bex, G.J., Neven, F., Schwentick, T., Tuyls, K.: Inference of concise DTDs from XML data. In: VLDB, pp. 115–126 (2006)

[Bex et al., 2007] Bex, G.J., Neven, F., Vansummeren, S.: Inferring XML schema definitions from XML data. In: VLDB, pp. 998–1009 (2007)

[Bouchou et al., 2009] Bouchou, B., Duarte, D., Halfeld Ferrari, M., Musicante, M.A.: Extending XML Types Using Updates. In: Hung, D. (ed.) Services and Business Computing Solutions with XML: Applications for Quality Management and Best Processes, pp. 1–21. IGI Global (2009)

[Bouchou et al., 2012] Bouchou, B., Halfeld Ferrari Alves, M., de Lima, M.A.V.: A grammarware for the incremental validation of integrity constraints on xml documents under multiple updates. T. Large-Scale Data- and Knowledge-Centered Systems 6, 167–197 (2012)

[Chabin et al.,] Chabin, J., Halfeld Ferrari, M., Musicante, M.A., Réty, P.: A software to transform a RTG into a LTG or a STTG., http://www.univ-orleans.fr/lifo/Members/rety/logiciels/RTGalgorithms.html

[Chabin et al., 2010] Chabin, J., Halfeld-Ferrari, M., Musicante, M.A., Réty, P.: Minimal Tree Language Extensions: A Keystone of XML Type Compatibility and Evolution. In: Cavalcanti, A., Deharbe, D., Gaudel, M.-C., Woodcock, J. (eds.) ICTAC 2010. LNCS, vol. 6255, pp. 60–75. Springer, Heidelberg (2010)

[Crane, 2006] Crane, G.: What do you do with a million books? D-Lib Magazine 12(3) (2006)

[da Luz et al., 2007] da Luz, R., Halfeld Ferrari, M., Musicante, M.A.: Regular expression transformations to extend regular languages (with application to a datalog XML schema validator). Journal of Algorithms 62(3-4), 148–167 (2007)

[Garofalakis et al., 2000] Garofalakis, M.N., Gionis, A., Rastogi, R., Seshadri, S., Shim, K.: Xtract: A system for extracting document type descriptors from xml documents. In: SIGMOD Conference, pp. 165–176 (2000)

[Gelade et al., 2010] Gelade, W., Idziaszek, T., Martens, W., Neven, F.: Simplifying xml schema: single-type approximations of regular tree languages. In: ACM SIGMOD-SIGACT-SIGART Symposium on Principles of Database Systems, PODS, pp. 251–260 (2010)

[Genevès et al., 2009] Genevès, P., Layaïda, N., Quint, V.: Identifying query incompatibilities with evolving xml schemas. SIGPLAN Not. 44, 221–230 (2009)

[Gold, 1967] Gold, E.M.: Language identification in the limit. Information and Control 10(5), 447–474 (1967)

[Gu et al., 2008] Gu, J., Xu, B., Chen, X.: An XML query rewriting mechanism with multiple ontologies integration based on complex semantic mapping. Information Fusion 9(4), 512–522 (2008)

[Guerrini et al., 2005] Guerrini, G., Mesiti, M., Rossi, D.: Impact of XML schema evolution on valid documents. In: WIDM 2005: Proceedings of the 7th Annual ACM International Workshop on Web Information and Data Management, pp. 39–44. ACM Press, New York (2005)

[Lambrix et al., 2008] Lambrix, P., Tan, H., Liu, Q.: Sambo and sambodtf results for the ontology alignment evaluation initiative 2008. In: OM (2008)

[Mani and Lee, 2002] Mani, M., Lee, D.: XML to Relational Conversion using Theory of Regular Tree Grammars. In: Bressan, S., Chaudhri, A.B., Li Lee, M., Yu, J.X., Lacroix, Z. (eds.) EEXTT and DIWeb 2002. LNCS, vol. 2590, pp. 81–103. Springer, Heidelberg (2003)

[Martens et al., 2006] Martens, W., Neven, F., Schwentick, T., Bex, G.J.: Expressiveness and complexity of XML schema. ACM Trans. Database Syst. 31(3), 770–813 (2006)

[Maßmann et al., 2011] Maßmann, S., Raunich, S., Aumüller, D., Arnold, P., Rahm, E.: Evolution of the coma match system. In: OM (2011)

[Moro et al., 2007] Moro, M.M., Malaika, S., Lim, L.: Preserving xml queries during schema evolution. In: Proceedings of the 16th International Conference on World Wide Web, WWW 2007, pp. 1341–1342. ACM (2007)

[Murata et al., 2005] Murata, M., Lee, D., Mani, M., Kawaguchi, K.: Taxonomy of XML schema languages using formal language theory. ACM Trans. Inter. Tech. 5(4), 660–704 (2005)

[Papakonstantinou and Vianu, 2000] Papakonstantinou, Y., Vianu, V.: DTD inference for views of XML data. In: PODS-Symposium on Principles of Database System, pp. 35–46. ACM Press (2000)

[Sakakibara, 1997] Sakakibara, Y.: Recent advances of grammatical inference. Theor. Comput. Sci. 185(1), 15–45 (1997)

[Shvaiko and Euzenat, 2005] Shvaiko, P., Euzenat, J.: A survey of schema-based matching approaches. In: Spaccapietra, S. (ed.) Journal on Data Semantics IV. LNCS, vol. 3730, pp. 146–171. Springer, Heidelberg (2005)

[Tekli et al., 2011] Tekli, J., Chbeir, R., Traina, A.J.M., Traina, C.: XML document-grammar comparison: related problems and applications. Central European Journal of Computer Science 1(1), 117–136 (2011)

[van den Brand et al., 2001] van den Brand, M., Heering, J., de Jong, H., de Jonge, M., Kuipers, T., Klint, P., Moonen, L., Olivier, P., Scheerder, J., Vinju, J., Visser, E., Visser, J.: The ASF+SDF meta-environment: a component-based language development environment. Electronic Notes in Theoretical Computer Science 44(2) (2001)

[Wu et al., 2001] Wu, X., Ling, T.W., Lee, M.-L., Dobbie, G.: Designing semistructured databases using ORA-SS model. In: Proceedings of the 2nd International Conference on Web Information Systems Engineering, WISE, (1) (2001)

A   Appendix: Proof of Theorem 1

We start by proving that G (the grammar obtained by Algorithm 1) is an LTG. Then we show that Algorithm 1 proposes a grammar which generates a language containing $L(G_0)$.

Lemma 5. *G is an LTG.*   □
Proof : By contradiction. Suppose $\hat{A} = \{A_1, ..., A_n\}$ and $\hat{B} = \{B_1, ..., B_k\}$ are competing in G, we have $\hat{A} \to a[R]$ and $\hat{B} \to a[R']$ in P. By construction of the rules of G (Algorithm 1) we must have in P_0 the following rules: $A_1 \to a[R_1], \cdots, A_n \to a[R_n], B_1 \to a[R'_1], \cdots, B_k \to a[R'_k]$. We deduce that $A_1, \cdots, A_n, B_1, \cdots, B_k$ are competing in G_0. Thus $\hat{A} = \hat{B}$. This is impossible since by definition, competing non-terminals are different. As, by construction, there is one rule in P for each element of $N = N_0/_{\parallel}$, G is in normal form.   □

The following lemma shows that the algorithm preserves the trees generated by the grammar G_0.

Lemma 6. *If $X \to^*_{G_0} t_0$ then $\hat{X} \to^*_G t_0$.*   □

Proof : By induction on the length of $X \to^*_{G_0} t_0 = a(w)$. Let us consider the first step of the derivation $X \to_{G_0} a[R_X]$ and $\exists U \in L(R_X), U \to^*_{G_0} w$. By construction of G, $\hat{X}$ is in N and $\hat{X} \to a[R]$ is in P with $R = \hat{R}_1 \mid \cdots \mid \hat{R}_n$, R_X is one of R_i so $\hat{U} \in L(R)$. Then $\hat{X} \to_G a[R]$ and by induction hypothesis, $\hat{U} \to^*_G w$ therefore $\hat{X} \to^*_G a(w) = t_0$.   □

We can now begin to show the relationship between the original language, as described by the grammar G_0 and the language generated by the grammar G obtained by the Algorithm 1.

Lemma 7. *(A) $L(G_0) \subseteq L(G)$. (B) If $X \to^*_{G_0} t$ and $\hat{X} \to^*_G t'$ then $t(\epsilon) = t'(\epsilon)$ (i.e. t and t' have the same top symbol).*   □

Proof : Item **(A)** is an immediate consequence of Lemma 6. As G_0 is in normal form, we have only one rule in G_0 of the form $X \to a[R_X]$, *i.e.*, the terms generated by X in G_0 have a as top symbol. Non-terminal X is in $\hat{X}$. From Lemma 5, we know that $\hat{X} \to a[R]$ is the unique rule in G whose left-hand side is $\hat{X}$.   □

From Example 4, we can, for instance, derive $Instrs \to^*_G ins(step, step)$ and $Instrs \to^*_G ins(number, step)$. Different terms with the same top label. Now, the next lemma states that the resulting grammar G does not introduce any new terminal symbol at the root of the generated trees.

Lemma 8. *$\forall t \in L(G), \exists t' \in L(G_0)$ such that $t'(\epsilon) = t(\epsilon)$.*

Proof : Let $t = a(w) \in L(G)$. Then there exists a rule $\{A_1, \cdots, A_n\} \to a[R]$ in P with $\{A_1, \cdots, A_n\} \in S$ (a start symbol in G). By definition of S, $\exists i$ such that $A_i \in S_0$ and $A_i \to^*_{G_0} a(w') = t'$ because G_0 is in the reduced form. So $t' \in L(G_0)$ and t and t' have the same root symbol.   □

Next, we show that for every subtree generated by G, its root appears in at least one subtree of the language generated by G_0 (recall that $w'(\epsilon) = w(\epsilon)$ means that forests w' and w have the same tuple of top-symbols):

Lemma 9. *If* $t \in ST(L(G))$, *such that* $t = a(w)$, *then,* $\exists t' \in ST(L(G_0))$, $t' = a(w') \wedge w'(\epsilon) = w(\epsilon)$.

Proof : Let $t \in ST(L(G))$ such that $t = a(w)$. There exists $\hat{A}_1 \rightarrow a[R] \in P$ such that $\hat{A}_1 \rightarrow a(U)$, $U \in L(R)$, $U \rightarrow_G^* w$. By construction, $R = \hat{R}_1 | \ldots | \hat{R}_n$, and $\forall i$, $\exists A_i \in N_0$, $A_i \rightarrow a[R_i] \in P_0 \wedge A_i \in \hat{A}_1$. Therefore there exists j such that $U \in L(\hat{R}_j)$. Consider $A_j \rightarrow a[R_j] \in P_0$. There exists $U' \in L(R_j)$ such that $\hat{U}' = U$. Now, since G_0 is in reduced form, there exists a forest w' such that $U' \rightarrow_{G_0}^* w'$. Consequently $A_j \rightarrow_{G_0} a(U') \rightarrow_{G_0}^* a(w')$. Let $t' = a(w')$. Since G_0 is in reduced form, the rule $A_j \rightarrow a[R_j]$ is reachable in G_0, then $t' \in ST(L(G_0))$. From Lemma 7, since $\hat{U}' = U$, we have $w(\epsilon) = w'(\epsilon)$. □

As an illustration of Lemma 9, we observe that from the grammars of Example 4, given the tree $r(is(ing(name,unit,qty)), r(req(item),ing(name,unit,qty), ins(step,step)), ins(step))$ from $L(G)$, for its sub-tree $t = r(req(item),ing(name, unit,qty),ins(step,step)) = r(w) \in ST(L(G))$, we have $t' = r(req(item),ing(name, qty,unit),ins(number,step)) = r(w') \in ST(L(G_0))$, $t(\epsilon) = t'(\epsilon)$ and $w(\epsilon) = w'(\epsilon)$.

Now we need some properties of local tree languages. The following lemma states that the type of the subtrees of a tree node is determined by the label of its node (i.e. the type of each node is *locally* defined). Recall that $ST(L)$ is the set of sub-trees of elements of L.

Lemma 10 (See [Papakonstantinou and Vianu, 2000] (Lemma 2.10)).
Let L *be a local tree language (LTL). Then, for each* $t \in ST(L)$, *each* $t' \in L$ *and each* $p' \in Pos(t')$, *we have that :* $t(\epsilon) = t'(p') \implies t'[p' \leftarrow t] \in L$. □

We also need a weaker version of the previous lemma:

Corollary 1. *Let* L *be a local tree language (LTL). Then, for each* $t, t' \in ST(L)$, *and each* $p' \in Pos(t')$, *we have that :* $t(\epsilon) = t'(p') \implies t'[p' \leftarrow t] \in ST(L)$. □

In practical terms, Corollary 1 gives us a rule of thumb on how to "complete" a regular language in order to obtain a local tree language. For instance, let $L = \{f(a(b),c), f(a(c),b)\}$ be a regular language. According to Corollary 1, we know that L is not LTL and that the least local tree language L' containing L contains all trees where a has c as a child together with all trees where a has b as a child. In other words, $L' = \{f(a(b),c), f(a(c),b), f(a(c),c), f(a(b),b)\}$.

Now, we can prove that the algorithm just adds what is necessary to get an LTL (and not more), in other words, that $L(G)$ is the least local tree language that contains $L(G_0)$. This is done in two stages: first for subtrees, then for trees.

Lemma 11. *Let* L' *be an LTL such that* $L(G_0) \subseteq L'$. *Then* $ST(L(G)) \subseteq ST(L')$.

Proof : By structural induction on the trees in $ST(L(G))$. Let $t = a(w) \in ST(L(G))$. From Lemma 9, there exists $t' \in ST(L(G_0))$ such that $t' = a(w') \wedge w'(\epsilon) = w(\epsilon)$. Since $L(G_0) \subseteq L'$, we have $ST(L(G_0)) \subseteq ST(L')$, then $t' \in ST(L')$. If w is an empty forest (i.e. the empty tuple), w' is also empty, therefore $t = t' \in ST(L')$. Otherwise, let us write $w = (a_1(w_1), \ldots, a_n(w_n))$ and $w' =$

$(a_1(w_1'), \ldots, a_n(w_n'))$ (since $w(\epsilon) = w'(\epsilon)$, w and w' have the same top symbols). Since $t = a(w) \in ST(L(G))$, for each $j \in \{1, \ldots, n\}$, $a_j(w_j) \in ST(L(G))$, then by induction hypothesis $a_j(w_j) \in ST(L')$. L' is an LTL, and for each j we have : $a_j(w_j) \in ST(L')$, $t' \in ST(L')$, $(a_j(w_j))(\epsilon) = a_j = t'(j)$. By applying Corollary 1 n times, we get $t'[1 \leftarrow a_1(w_1)] \ldots [n \leftarrow a_n(w_n)] = t \in ST(L')$. $\qquad \square$

Theorem 4. *Let L' be an LTL such that $L' \supseteq L(G_0)$. Then $L(G) \subseteq L'$.*

Proof : Let $t \in L(G)$. Then $t \in ST(L(G))$. From Lemma 11, $t \in ST(L')$. On the other hand, from Lemma 8, there exists $t' \in L(G_0)$ such that $t'(\epsilon) = t(\epsilon)$. Then $t' \in L'$. From Lemma 10, $t'[\epsilon \leftarrow t] = t \in L'$. $\qquad \square$

This result ensures that the grammar G of Example 4 generates the least LTL that contains $L(G_0)$.

B   Appendix: Proof of Theorem 2

The proof somehow looks like the proof concerning the transformation of an RTG into an LTG (Section 4). However it is more complicate since in a STTL (and unlike what happens in an LTL), the confusion between $t|_p = a(w)$ and $t'|_{p'} = a(w')$ should be done only if position p in t has been generated by the same production rule as position p' in t', i.e. the symbols occurring in t and t' along the paths going from root to p (resp. p' in t') are the same. This is why we introduce notation $path(t, p)$ to denote these symbols (Definition 7).

Lemma 12. *Let $\chi \in \mathcal{P}(N_0)$ and $A, B \in \chi$. Then $\hat{A}^\chi$ and $\hat{B}^\chi$ are not competing in P.* $\qquad \square$

Proof: By contradiction. Suppose $\hat{A}^\chi$ and $\hat{B}^\chi$ are competing in P. Then there exist $\hat{A}^\chi \to a[R_1] \in P$ and $\hat{B}^\chi \to a[R_2] \in P$. From the construction of P, there exist $C \in \hat{A}^\chi$ (then $C \parallel_\chi A$) and $C \to a[R_1'] \in P_0$, as well as $D \in \hat{B}^\chi$ (then $D \parallel_\chi B$) and $D \to a[R_2'] \in P_0$. Thus, $C \parallel_\chi D$ and by transitivity $A \parallel_\chi B$, then $\hat{A}^\chi = \hat{B}^\chi$, which is impossible since competing non-terminals are not equal. $\qquad \square$

Example 9. Consider the grammar of Example 6.

Let $\chi = \{Frame1, Frame2, Background\}$. The equivalence classes induced by $\parallel_\chi$ are $\widehat{Frame1}^\chi = \widehat{Frame2}^\chi = \{Frame1, Frame2\}$; $\widehat{Background}^\chi = \{Background\}$; which are non-competing non-terminals in P. $\qquad \square$

Lemma 13. *$G = (N, T, S, P)$ is a STTG.* $\qquad \square$

Proof: (1) There is no regular expression in P containing competing non-terminals: If $\hat{A}^{S_0}, \hat{B}^{S_0}$ are in S, then $A, B \in S_0$. From Lemma 12, $\hat{A}^{S_0}$ and $\hat{B}^{S_0}$ are not competing in P. For any regular expression R, let $\hat{A}^{N(R)}, \hat{B}^{N(R)} \in N(\sigma_{N(R)}(R))$. Thus, $A, B \in N(R)$. From Lemma 12, $\hat{A}^{N(R)}$ and $\hat{B}^{N(R)}$ are not competing in P. (2) G is in normal form: As for each A_i there is at most one rule in P_0 whose left-hand-side is A_i (because G_0 is in normal form), there is at most one rule in P whose left-hand-side is $\{A_1, \ldots, A_n\}$. $\qquad \square$

The next lemma establishes the basis for proving that the language generated by G contains the language generated by G_0. It considers the derivation process over G_0 at any step (supposing that this step is represented by a derivation tree t) and proves that, in this case, at the same derivation step over G, we can obtain a tree t' having all the following properties: (i) the set of positions is the same for both trees $(Pos(t) = Pos(t'))$; (ii) positions associated to terminal are identical in both trees; (iii) if position p is associated to a non-terminal A in t then position $p \in Pos(t')$ is associated to the equivalence class $\hat{A}^\chi$ for some $\chi \in \mathcal{P}(N_0)$ such that $A \in \chi$.

Lemma 14. *Let $Y \in S_0$. If G_0 derives:*
$t_0 = Y \to \cdots \to t_{n-1} \to_{[p_n, A_n \to a_n[R_n]]} t_n$ *then G can derive:* $t'_0 = \hat{Y}^{S_0} \to \cdots \to$
$t'_{n-1} \to_{[p_n, \hat{A}_n{}^{\chi_n} \to a_n[\sigma_{N(R_n|\cdots)}(R_n|\cdots)]]} t'_n$
s.t. $\forall i \in \{0, \ldots, n\}, Pos(t'_i) = Pos(t_i) \wedge$
$\forall p \in Pos(t_i): (t_i(p) \in T_0 \implies t'_i(p) = t_i(p)) \wedge$
$$(t_i(p) = A \in N_0 \implies \exists \chi \in \mathcal{P}(N_0),\, A \in \chi \wedge t'_i(p) = \hat{A}^\chi) \qquad \square$$

Proof: See [Chabin et al., 2010].

The following corollary proves that the language of the new grammar G, proposed by Algorithm 2, contains the original language of G_0.

Corollary 2. $L(G_0) \subseteq L(G)$. $\qquad\qquad\square$

In the rest of this section we work on proving that $L(G)$ is the least STTL that contains $L(G_0)$. To prove this property, we first need to prove some properties over STTLs. We start by considering paths in a tree. We are interested by paths (sequence of labels) starting on the root and achieving a given position p in a tree t. For example, $path(a(b, c(d)), 1) = a.c$.

Definition 7 (Path in a tree t to a position p). *Let t be a tree and $p \in Pos(t)$, then $path(t, p)$ is recursively defined by:* (1) $path(t, \epsilon) = t(\epsilon)$ *and* (2) $path(t, p.i) = path(t, p).t(p.i)$ *where $i \in \mathbb{N}$.* $\qquad\square$

Given a STTG G, let us consider the derivation process of two trees t and t' belonging to $L(G)$. The following lemma proves that positions (p in t and p' in t') having identical paths are derived by using the same rules. A consequence of this lemma (when $t' = t$ and $p' = p$) is the well known result about the unicity in the way of deriving a given tree with a STTG [Mani and Lee, 2002].

Lemma 15. *Let G' be a STTG, let $t, t' \in L(G')$. Let $X \to^*_{[p_i, rule_{p_i}]} t$ be a derivation of t and $X' \to^*_{[p'_i, rule'_{p'_i}]} t'$ be a derivation of t' by G' (X, X' are start symbols). Then $\forall p \in Pos(t), \forall p' \in Pos(t'), (path(t, p) = path(t', p') \implies rule_p = rule'_{p'})$* $\qquad\square$

Proof: Suppose $path(t, p) = path(t', p')$. Then we have $length(p) = length(p')$. The proof is by induction on $length(p)$.

- If $length(p) = 0$, then $p = p' = \epsilon$, and $t(\epsilon) = t'(\epsilon) = a$. Therefore $rule_\epsilon = (X \to a[R])$ and $rule'_\epsilon = (X' \to a[R'])$. If $X \neq X'$ then two start symbols are competing, which is impossible since G' is a STTG. If $X = X'$ and $R \neq R'$ then G' is not in normal form, which contradicts the fact that G' is a STTG. Therefore $rule_\epsilon = rule'_\epsilon$, then $rule_p = rule'_{p'}$.

• Induction step. Suppose $p = q.k$ and $p' = q'.k'$ ($k, k' \in \mathbb{N}$), and $path(t, p) = path(t', p')$. Then $path(t, q) = path(t', q')$. By induction hypothesis, $rule_q = rule'_{q'} = (X \rightarrow a[R])$. There exits $w, w' \in L(R)$ s.t. $w(k) = A$, $w'(k') = A'$ and $rule_p = (A \rightarrow b[R_1])$, $rule'_{p'} = (A' \rightarrow b[R'_1])$ where $b = t(p) = t'(p')$. If $A \neq A'$ then $A \in N(R)$ and $A' \in N(R)$ are competing, which is impossible since G' is a STTG. If $A = A'$ and $R_1 \neq R'_1$, then G' is not in normal form, which contradicts the fact that G' is a STTG. Consequently $rule_p = rule'_{p'}$. □

In a STTL, it is possible to permute sub-trees that have the same paths.

Lemma 16 (Also in [Martens et al., 2006]). *Let G' be a STTG. Then,* $\forall t, t' \in L(G')$, $\forall p \in Pos(t)$, $\forall p' \in Pos(t')$, $(path(t, p) = path(t', p') \implies t'[p' \leftarrow t|_p] \in L(G'))$. □

Example 10. Let G be the grammar of Example 6. Consider a tree t as shown in Figure 3. The permutation of subtrees $t|_{0.0}$ and $t|_{0.1}$ gives us a new tree t'. Both t and t' are in $L(G)$. □

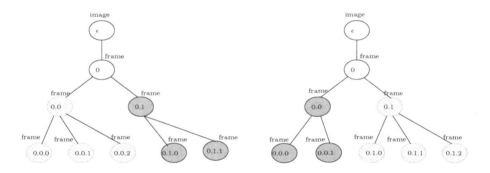

Fig. 3. Trees t and t' with permuted sub-trees

Definition 8 (Branch Derivation). *Let G' be an RTG. A branch-derivation is a tuple of production rules[7] of G': $\langle A^1 \rightarrow a^1[R^1], \ldots, A^n \rightarrow a^n[R^n] \rangle$ s.t. $\forall i \in \{2, \ldots, n\}, A^i \in N(R^{i-1})$.* □

Notice that if A^1 is a start symbol, the branch-derivation represents the derivation of a branch of a tree[8]. This branch contains the terminals $a^1, \ldots, a^n$ (the path to the node having a^n as label). Now, let us prove properties over the grammar G built by Algorithm 2.

---

[7] Indices are written as super-scripts for coherence with the notations in Lemma 17.
[8] This tree may contain non-terminals.

Lemma 17. *Consider a branch-derivation in G:*
$$\langle\{A_1^1,\ldots,A_{n_1}^1\} \to a^1\sigma_{N(R_1^1|\cdots|R_{n_1}^1)}[R_1^1|\cdots|R_{n_1}^1],\ldots,$$
$$\{A_1^k,\ldots,A_{n_k}^k\} \to a^k\sigma_{N(R_1^k|\cdots|R_{n_k}^k)}[R_1^k|\cdots|R_{n_k}^k]\rangle \text{ and let } i_k \in \{1,\ldots,n_k\}. \text{ Then}$$
there exists a branch-derivation in G_0: $(A_{i_1}^1 \to a^1[R_{i_1}^1],\ldots,A_{i_k}^k \to a^k[R_{i_k}^k])$. □

Proof: By induction on k.

- $k = 1$. There is one step. From Definition 2, $A_{i_k}^k \to a^k[R_{i_k}^k] \in P_0$.
- Induction step. By induction hypothesis applied on the last $k-1$ steps, there exists a branch-derivation in G_0 : $(A_{i_2}^2 \to a^2[R_{i_2}^2],\ldots,A_{i_k}^k \to a^k[R_{i_k}^k])$.
Moreover $\{A_1^2,\ldots,A_{n_2}^2\} \in N(\sigma_{N(R_1^1|\cdots|R_{n_1}^1)}(R_1^1|\cdots|R_{n_1}^1))$. Then there exists $i_1 \in \{1,\ldots,n_1\}$ s.t. $A_{i_2}^2 \in N(R_{i_1}^1)$. And from Definition 2, $A_{i_1}^1 \to a^1[R_{i_1}^1] \in P_0$. □

The following example illustrates Lemma 17 and its proof.

Example 11. Let G be the grammar of Example 6 and t the tree of Figure 3. The branch-derivation corresponding to the node 0.0.0 contains the first and the fourth rules of G presented in Example 6 (notice that the fourth rule appears three times). Figure 4 illustrates this branch-derivation on a derivation tree. For instance, the first rule in G is

$$\{Image\} \to image[\{Frame1, Frame2\} \mid \{Background\}.\{Foreground\}] \qquad (R1)$$

and G_0 has the production rule
$Image \to image[Frame1 \mid Frame2 \mid Background.Foreground]$. Then, the branch-derivation gives us the fourth rule in G, namely:

$$\{Frame1, Frame2\} \to frame[\epsilon$$
$$\mid \{Frame1, Frame2\}.\{Frame1, Frame2\}$$
$$\mid \{Frame1, Frame2\}.\{Frame1, Frame2\}.\{Frame1, Frame2\}].$$

Notice that the left-hand side $\{Frame1, Frame2\}$ is a non terminal in the right-hand side of (R1). Now, consider each non terminal of G_0 forming the non terminal $\{Frame1, Frame2\}$ in G. Clearly, $Frame1$ is on the right-hand side of the second rule in P_0 while $Frame2$ is on the right-hand side of the third rule in P_0 (as shown in Example 6). We can observe the same situation for all the rules in the branch-derivation. Thus, as proved in Lemma 17, the branch-derivation in G_0 that corresponds to the one considered in this example is:

$$\langle\ Image \to image[Frame1 \mid Frame2 \mid Background.Foreground]$$
$$Frame2 \to frame[Frame2.Frame2.Frame2 \mid \epsilon]$$
$$Frame2 \to frame[Frame2.Frame2.Frame2 \mid \epsilon]$$
$$Frame2 \to frame[Frame2.Frame2.Frame2 \mid \epsilon]\ \rangle \qquad\qquad □$$

The following lemma somehow expresses what the algorithm of Definition 2 does. Given a forest $w = (t_1,\ldots,t_n)$, recall that $w(\epsilon) = \langle t_1(\epsilon),\ldots,t_n(\epsilon)\rangle$, i.e. $w(\epsilon)$ is the tuple of the top symbols of w.

Lemma 18. $\forall t \in L(G), \forall p \in Pos(t),$
$t|_p = a(w) \implies \exists t' \in L(G_0), \exists p' \in pos(t'), t'|_{p'} = a(w') \wedge w'(\epsilon) = w(\epsilon) \wedge path(t',p') = path(t,p).$ □

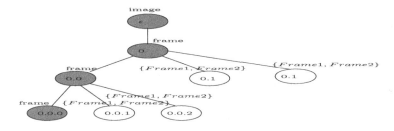

Fig. 4. Derivation tree in G. Grey nodes illustrate a branch-derivation.

Proof: There exists a branch-derivation in G that derives the position p of t
$$(\{A_1^1, \ldots, A_{n_1}^1\} \to a^1 \sigma_{N(R_1^1|\cdots|R_{n_1}^1)}[R_1^1| \cdots |R_{n_1}^1], \ldots,$$
$$\{A_1^k, \ldots, A_{n_k}^k\} \to a^k \sigma_{N(R_1^k|\cdots|R_{n_k}^k)}[R_1^k| \cdots |R_{n_k}^k])$$
and $u \in L(\sigma_{N(R_1^k|\cdots|R_{n_k}^k)}(R_1^k| \cdots |R_{n_k}^k))$ s.t. $u \to_G^* w$.

Then there exists i_k s.t. $u \in L(\sigma_{N(R_1^k|\cdots|R_{n_k}^k)}(R_{i_k}^k))$. Thus, there exists $v \in L(R_{i_k}^k)$
s.t. $u = \sigma_{N(R_1^k|\cdots|R_{n_k}^k)}(v)$. Note that $\forall Y \in N_0$, $\forall \chi \in \mathcal{P}(N_0)$, Y and $\hat{Y}^\chi$ generate
the same top-symbol. So u and v generate the same top-symbols. Since G_0 is in
reduced form, there exists w' s.t. $v \to_{G_0}^* w'$, and then $w'(\epsilon) = w(\epsilon)$.

From Lemma 17, there exists a branch-derivation in G_0: $(A_{i_1}^1 \to a^1[R_{i_1}^1], \ldots,$
$A_{i_k}^k \to a^k[R_{i_k}^k])$. Since G_0 is in reduced form, there exists $t' \in L_{G_0}(A_{i_1}^1)$ (i.e. t'
is a tree derived from $A_{i_1}^1$ by rules in P_0, and t' contains only terminals), and
there exists $p' \in Pos(t')$ s.t. this branch-derivation derives in G_0 the position
p' of t'. Since $v \in L(R_{i_k}^k)$ and $v \to_{G_0}^* w'$, one can even choose t' s.t. $t'|_{p'} =$
$a^k(w')$. Since $a^k = a$, we have $t'|_{p'} = a(w')$. On the other hand, $path(t', p') =$
$a^1 \ldots a^k = path(t, p)$. Finally, since $t \in L(G)$, $\{A_1^1, \ldots, A_{n_1}^1\} \in S$. Since $A_{i_1}^1 \in$
$\{A_1^1, \ldots, A_{n_1}^1\}$, from Definition 2 we have $A_{i_1}^1 \in S_0$. Therefore $t' \in L(G_0)$. □

Example 12. Let G be the grammar of Example 6 and t the tree of Figure 3.
Let $p = 0$. Using the notations of Lemma 18, $t|_0 = frame(w)$ where
$w = \langle frame(frame, frame, frame),\ frame(frame, frame) \rangle$. We have $t \notin$
$L(G_0)$. Let $t' = image(frame(frame(frame, frame),\ frame)) \in L(G_0)$ and
(with $p' = p = 0$) $t'|_{p'} = frame(w')$ where $w' = \langle frame(frame, frame),$
$frame \rangle$. Thus $w'(\epsilon) = w(\epsilon)$. Note that others $t' \in L(G_0)$ suit as well. □

We end this section by proving that the grammar obtained by our algorithm
generates the least STTL which contains $L(G_0)$.

Lemma 19. *Let L' be a STTL s.t. $L(G_0) \subseteq L'$. Let $t \in L(G)$. Then, $\forall p \in$
$Pos(t), \exists t' \in L', \exists p' \in pos(t'), (t'|_{p'} = t|_p \wedge path(t', p') = path(t, p))$.* □

Proof: We define the relation $\sqsupset$ over $Pos(t)$ by $p \sqsupset q \iff \exists i \in \mathbb{N}, p.i = q$. Since
$Pos(t)$ is finite, $\sqsupset$ is noetherian. The proof is by noetherian induction on $\sqsupset$. Let
$p \in pos(t)$. Let us write $t|_p = a(w)$.

From Lemma 18, we know that: $\exists t' \in L(G_0), \exists p' \in pos(t'), t'|_{p'} = a(w') \wedge w'(\epsilon) = w(\epsilon) \wedge path(t', p') = path(t, p)$. Thus, $t|_p = a(a_1(w_1), \ldots, a_n(w_n))$ and $t'|_{p'} = a(a_1(w'_1), \ldots, a_n(w'_n))$.

Now let $p \sqsupset p.1$. By induction hypothesis: $\exists t'_1 \in L', \exists p'_1 \in pos(t'_1), t'_1|_{p'_1} = t|_{p.1} = a_1(w_1) \wedge path(t'_1, p'_1) = path(t, p.1)$. Notice that $t'_1 \in L'$, $t' \in L(G_0) \subseteq L'$, and L' is a STTL. Moreover $path(t'_1, p'_1) = path(t, p.1) = path(t, p).a_1 = path(t', p').a_1 = path(t', p'.1)$.

As $path(t'_1, p'_1) = path(t', p'.1)$, from Lemma 16 applied on t'_1 and t', we get $t'[p'.1 \leftarrow t'_1|_{p'_1}] \in L'$. However $(t'[p'.1 \leftarrow t'_1|_{p'_1}])|_{p'} = a(a_1(w_1), a_2(w'_2), \ldots, a_n(w'_n))$ and $path(t'[p'.1 \leftarrow t'_1|_{p'_1}], p') = path(t', p') = path(t, p)$. By applying the same reasoning for positions $p.2, \ldots, p.n$, we get a tree $t'' \in L'$ such that $t''|_{p'} = t|_p$ and $path(t'', p') = path(t, p)$. $\qquad\square$

Corollary 3 (When $p = \epsilon$, and then $p' = \epsilon$). *Let L' be a STTL such that $L' \supseteq L(G_0)$. Then $L(G) \subseteq L'$.* $\qquad\square$

Pairwise Similarity
for Cluster Ensemble Problem:
Link-Based and Approximate Approaches

Natthakan Iam-On[1,*] and Tossapon Boongoen[2]

[1] School of Information Technology, Mae Fah Luang University,
Chiang Rai 57100, Thailand
`nt.iamon@gmail.com`
[2] Department of Mathematics and Computer Science,
Royal Thai Air Force Academy,
Bangkok 10220, Thailand
`tossapon_b@rtaf.mi.th`

Abstract. Cluster ensemble methods have emerged as powerful techniques, aggregating several input data clusterings to generate a single output clustering, with improved robustness and stability. In particular, link-based similarity techniques have recently been introduced with superior performance to the conventional co-association method. Their potential and applicability are, however limited due to the underlying time complexity. In light of such shortcoming, this paper presents two approximate approaches that mitigate the problem of time complexity: the approximate algorithm approach (Approximate SimRank Based Similarity matrix) and the approximate data approach (Prototype-based cluster ensemble model). The first approach involves decreasing the computational requirement of the existing link-based technique; the second reduces the size of the problem by finding a smaller, representative, approximate dataset, derived by a density-biased sampling technique. The advantages of both approximate approaches are empirically demonstrated over 22 datasets (both artificial and real data) and statistical comparisons of performance (with 95% confidence level) with three well-known validity criteria. Results obtained from these experiments suggest that approximate techniques can efficiently help scaling up the application of link-based similarity methods to wider range of data sizes.

Keywords: clustering, cluster ensembles, pairwise similarity matrix, cluster relation, link analysis, data prototype.

1   Introduction

Data clustering is a very common task, playing a crucial role in a number of application domains, such as machine learning, data mining, information retrieval

---

*   Corresponding author.

A. Hameurlain et al. (Eds.): TLDKS IX, LNCS 7980, pp. 95–122, 2013.
© Springer-Verlag Berlin Heidelberg 2013

and pattern recognition. Clustering aims to categorize data into groups or clusters such that the data in the same cluster are more similar to each other than to those in different clusters, with the underlying structure of real-world datasets containing a bewildering combination of shape, size and density. Although, a large number of clustering algorithms have been introduced for a variety of application areas [18], the No Free Lunch theorem [36] suggests there is no single clustering algorithm that performs best for all datasets [25], i.e. unable to discover all types of cluster shapes and structures presented in data [7], [13], [37]. Each algorithm has its own strengths and weaknesses. For any given dataset, it is usual for different algorithms to provide distinct solutions; indeed, apparent structural differences may occur within the same algorithm, given different parameters. As a result, it is extremely difficult for users to decide a priori which algorithm would be the *the most appropriate* for a given set of data.

Recently, the cluster ensemble approach has emerged as an effective solution that is able to overcome these problems; moreover, it improves robustness, as well as the quality of clustering results. The main objective of the cluster ensemble approach is to combine the different decisions of various clustering algorithms in such a way as to achieve an accuracy superior to individual clusterings. Examples of well-known ensemble methods are: (i) the feature-based approach that transforms the problem of cluster ensembles to clustering categorical data [34], [35], (ii) graph-based algorithms that employ a graph partitioning methodology [32], and (iii) the pairwise similarity approach that makes use of co-occurrence relationships between all pairs of data points [13], [32].

Of particular interest here is the pairwise similarity approach, in which the final partitions are derived based on relations amongst data points represented within the similarity matrix. This is widely known as the *Co-Association* matrix [13]. This relation-oriented matrix denotes co-occurrence statistics between each pair of data points, especially in term of the proportion of base clusterings in which they are assigned to the same cluster. In essence, the co-association matrix can be regarded as a new similarity matrix, which is superior to the original distance based counterpart [17]. It has been wildly applied to various application domains such as gene expression data analysis [28], [33] and satellite image analysis [27].

This approach has gained popularity and become a practical alternative mainly due to its simplicity. However, it has been criticized because the underlying matrix only considers the similarity of data points at coarse level and completely ignores those existing amongst clusters [9], [16]. As a result, by not exploiting available information regarding cluster associations, many relations are left *unknown* with zero similarity value. To this extend, Iam-on et al. [16] introduced methods for generating two new pairwise similarity matrices, named *Connected-Triple Based Similarity* and *SimRank Based Similarity* matrices. Both are informed by the basic conjecture of taking into consideration as much information, embedded in a cluster ensemble, as possible when finding similarity between data points. To discover similarity values, they consider both the associations among data points as well as those among clusters in the ensemble using link-based similarity measures [4], [19], [23].

This paper aims to generalize the characteristics and performance of different pairwise similarity methods proposed in the literature for the cluster ensemble problem. To this end, the quality of clustering results achieved with these methods over both real and artificial datasets, of distinct shapes and sizes, are extensively examined. Furthermore, difficulties in their applicability are explored, especially the problem of high computational complexity of $O(N^2)$ (N being the number of data points). This paper describes how pair-wise techniques can be made more efficient, especially when applied to a large dataset, through approximate schemes: (i) reducing the complexity of link-based similarity estimation and (ii) reducing the number of data points, where a set of P ($P << N$) representative data points (i.e. prototypes) are exploited instead. Following that, a decision-support matrix is suggested with appropriate alternatives of pairwise methods and prototyping techniques for distinct requirement contexts, regarding time consumption, accuracy level and size of data.

The paper is organized as follows. Section 2 contains a formal definition of the cluster ensemble problem and its general framework. Section 3 presents a review on the pairwise similarity approach, including co-association and link-based similarity matrices. Approximate approaches to cluster ensemble problem and their underlying intuitions are thoroughly expressed in Section 4. Next, Section 5 generalizes the quality of these methods through their experimental evaluation under a variety of conditions and datasets. In Section 6, a decision-support matrix is introduced as the guideline to selecting appropriate pairwise similarity methods and approximation scheme for different requirement criteria. The paper is concluded in Section 7 with suggestions for further work.

2   Cluster Ensemble Problem

This section includes fundamental concepts regarding the problem of cluster ensemble upon which this research has been developed.

2.1   Problem Formulation

Let $X = \{x_1, \ldots, x_N\}$ be a set of N data points and let $\Pi = \{\pi_1, \ldots, \pi_M\}$ be a set of M base clustering results, which will be referred to as a *cluster ensemble*. Each base clustering result (called an *ensemble member*) returns a set of clusters $\pi_i = \{C_1^i, C_2^i, \ldots, C_{k_i}^i\}$, such that $\bigcup_{j=1}^{k_i} C_j^i = X$, where k_i is the number of clusters in the i-th clustering. For each $x \in X$, $C^i(x)$ denotes the cluster label to which the data point x belongs. In the i-th clustering, $C^i(x) = j$ if $x \in C_j^i$. The problem is to find a new partition π^* of a dataset X that summarizes the information from the cluster ensemble Π.

2.2   Cluster Ensemble Framework

The general process of cluster ensemble is shown in Figure 1. In particular, solutions achieved from different base clusterings, termed as *ensemble members*, are intelligently aggregated to form a final partition. Essentially, this meta-level method involves two major tasks of: (i) generating a cluster ensemble, and (ii) producing the final partition (normally referred to as *consensus function*). At the outset, a cluster ensemble is typically built by exploiting different cluster models and different data partitions. Distinct cluster models refer to different cluster algorithms or a single algorithm with several sets of parameter initialization, such as cluster centers and number of clusters used in k-means method [11], [12], [13], [15], [35]. In fact, a cluster ensemble can also be achieved by applying manifold subsets of initial data to base clusterings. It is intuitively assumed that each clustering algorithm can provide different levels of performance for different partitions of a dataset [6]. Practically, data partitions are obtained through either data projection (i.e. subspace) in which partitions possess identical number of data points but each with different collection of attributes, or data sampling where partitions are similarly characterized by all initial attributes but each with reduced number of data points [8], [34]. In addition to using one of these methods, any combination of them can be applied as well [6], [14], [28], [29], [32].

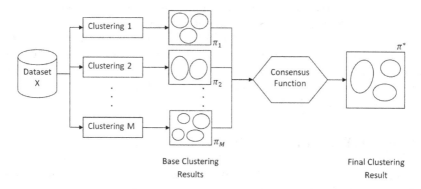

Fig. 1. The basic process of cluster ensembles. It first applies multiple base clusterings to a dataset X to obtain diverse clustering decisions ($\pi_1 \ldots \pi_M$). Then, these solutions are combined to establish the final clustering result (π^*) using a consensus function.

Having obtained the cluster ensemble, a variety of *consensus functions* (i.e. *methods*) have been developed and made available for deriving the ultimate data partition. In general, consensus methods can be categorized into: (i) feature based, (ii) graph based and (iii) pairwise similarity approaches. The first technique transforms the problem of cluster ensembles to clustering categorical data. Specifically, each base clustering provides a class label as a new feature describing each data point, which is utilized to formulate the ultimate solution [3], [15], [29], [34], [35]. The second methodology makes use of the graph representation to solve the cluster ensemble problem [6], [9], [32]. In essence, a graph representing an ensemble is divided into a definite number of approximately equal-sized partitions, using graph partitioning techniques like METIS [21] and HMETIS [20]. The last approach creates a matrix, containing the pairwise similarity among data points, to which any similarity-based clustering algorithms can be applied [8], [10], [11], [12], [13], [28], [29], [32].

3 Pairwise Cluster Ensemble

3.1 Co-association Method (Benchmark Method)

This specific category of cluster ensemble method employs the pairwise similarity approach as its consensus function. In particular, given a dataset $X = \{x_1, x_2, \ldots, x_N\}$, it first generates a cluster ensemble $\Pi = \{\pi_1, \pi_2, \ldots, \pi_M\}$ by applying M base clusterings to the dataset X. Following that, an $N \times N$ similarity matrix is constructed for each ensemble member, denoted as $S_m, m = 1 \ldots M$. Each entry in this matrix represents the relationship between two data points. If they are assigned to the same cluster, the entry will be 1, 0 otherwise. More precisely, the similarity between two data points x_i and x_j from the m-th ensemble member can be computed as follows:

$$S_m(x_i, x_j) = \begin{cases} 1 & if\, C^m(x_i) = C^m(x_j) \\ 0 & otherwise \end{cases} \tag{1}$$

In essence, M similarity matrices are merged to form a *co-association matrix (CO)* [13], various names found in the literature as consensus matrix [28], similarity matrix [32] or agreement matrix [33]. The elements in the CO matrix represent similarity between any two data points, which is a ratio of a number of ensemble members in which these data points are assigned to the same cluster to the total number of ensemble members. Formally, the similarity between two data points x_i and x_j across all base clusterings is defined as,

$$CO(x_i, x_j) = \frac{1}{M} \sum_{m=1}^{M} S_m(x_i, x_j) \tag{2}$$

Since the CO matrix is a similarity matrix, any similarity-based clustering algorithm can be applied on this matrix to yield the final partition π^*. Among several existing similarity-based methods, the most two well-known techniques are agglomerative clustering algorithm and graph partitioning method. Specifically, Fred and Jain [13], [12] and Monti et al. [28] made use of the agglomerative clustering to derive the final partitions. In contrary, Strehl and Ghosh [32] proposed Cluster-based Similarity Partitioning Algorithm (CSPA) that generates a similarity graph whose vertices represent data points and edges' weights represent similarity scores obtained from the CO matrix. Afterwards, a graph partitioning algorithm called METIS [21] is used to divide this similarity graph into k clusters of approximately equal size.

3.2 Link-Based Methods

Despite the advantage of its simplicity, the CO matrix fails drastically to handle *unknown* relations between data points, whose similarity is zero. Investigations have shown zero-similarity occurs 75% of the time, +/- 5%, for the real-world datasets used in [16]. The CO matrix can expose only a small proportion of pairwise similarity between data points, which may be better discovered by bringing in additional information regarding similarity relations between clusters in an ensemble. To be concise, such a relation determines the similarity of any two clusters in questions, which can be estimated from a graph or link network representing the ensemble. Inspired by this idea, Iam-on et al. [16] employed link-based similarity measures to refine the evaluation of similarity values among data points: the Connected-Triple Based similarity (CTS) and the SimRank Based Similarity (SRS) matrices, respectively.

Figure 2 shows much lower percentages of unknown relations were achieved in the CTS and SRS link-based similarity matrices, compared to the CO matrix. This evidence suggests that link-based similarity measures can help discover implicit relationship amongst data points, which is not possible using the original co-occurrence statistical approach.

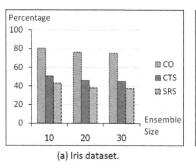
(a) Iris dataset.

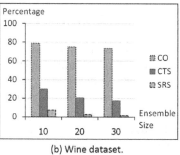
(b) Wine dataset.

Fig. 2. Percentages of zero-similarity values in two link-based similarity matrices, comparing to those of the CO matrix. This set of statistics is the average figures of 50 runs achieved on Iris and Wine datasets, with three different ensemble sizes (10, 20 and 30).

Connected-Triple Based Similarity (CTS) Matrix. The CTS matrix substantially extends the original CO matrix with the concept of cluster relations within a cluster ensemble. Strength-of-cluster associations can be quantitatively measured by the Connected-Triple approach, and this method has been used to disclose duplicate author names in the Digital Bibliography and Library Project (DBLP) database [23]. It works on the basis that if two nodes share a link to a third node then this is indicative of similarity between the two nodes. This principle is illustrated in Figure 3. The circled vertices denote data points and the square vertices represent clusters in each labelled clustering. There exists an edge between a data point x_i and a cluster C_j if x_i belongs to C_j. Note that data points x_1 and x_2 are considered to be similar in clusterings 2 and 3, in which they are assigned to the same clusters (clusters C and D, respectively). However, their similarity would be taken as zero using information in the clustering 1 alone. Intuitively, despite being assigned to different clusters, their similarity,

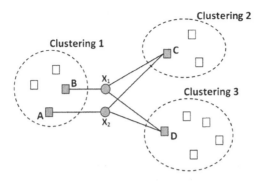

Fig. 3. A graphical representation of a cluster ensemble

and the partial similarity of clusters A and B, should be measurable. Using the Connected-Triple approach, this similarity becomes clear, and clusters A and B are shown to be similar (at least to some degree) due to the fact that they possess 2 Connected-Triples, with clusters C and D at the centers of the triples.

Originally, the number of triples associated with any two objects is summed. This simple counting might be sufficient for data points or other indivisible objects. However, to evaluate the similarity between clusters, it is crucial to take into account the characteristics like shared data members among clusters. Inspired by this idea, the new Weighted Connected-Triple algorithm for the problem of cluster ensembles was introduced [16] and is described as follows.

Weighted Connected-Triple. Given a cluster ensemble Π, a graph $G = (V, W)$ can be constructed where V is the set of vertices each representing a cluster in Π and W is a set of weighted edges between clusters. Formally, the weight assigned to the edge connecting clusters i and j is estimated in accordance with the proportion of their overlapping members.

$$w_{ij} = \frac{|X_i \cap X_j|}{|X_i \cup X_j|} \tag{3}$$

where X_A denotes the set of data points belonging to cluster A. Instead of counting the number of triples as a whole number, the Weighted Connected-Triple regards each triple as the minimum weight of the two involving edges.

$$C_{ij}^k = \min(w_{ik}, w_{jk}) \tag{4}$$

where C_{ij}^k is the count of the triple between clusters i and j whose common neighbor is cluster k. The count of all triples $(1 \ldots q)$ between cluster i and cluster j can be calculated as follows:

$$C_{ij} = \sum_{k=1}^{q} C_{ij}^k \tag{5}$$

The similarity between clusters i and j can be estimated as follows, where C_{max} is the maximum C_{ij} value of any two clusters i and j.

$$S_{WT}(i,j) = \frac{C_{ij}}{C_{max}} \tag{6}$$

Connected-Triple Based Similarity (CTS) Matrix. This matrix adopts the cluster-oriented approach previously described to enhance the quality of the pairwise similarity matrix. Specifically, for the m-th ensemble member, the similarity of data points x_i and x_j is estimated using the following equation, where DC is a constant decay factor (i.e. confidence level of accepting two non-identical objects, which has not been assigned to the same cluster, as being similar) whose value range is in $[0,1]$.

$$S_m(x_i, x_j) = \begin{cases} 1 & if\, C^m(x_i) = C^m(x_j) \\ S_{WT}(C^m(x_i), C^m(x_j)) \times DC & otherwise \end{cases} \tag{7}$$

Following that, each entry in the CTS matrix can be computed as,

$$CTS(x_i, x_j) = \frac{1}{M} \sum_{m=1}^{M} S_m(x_i, x_j) \tag{8}$$

SimRank Based Similarity (SRS) Matrix. This link-based matrix is built upon the SimRank algorithm [19] with the underlying assumption of *neighbors are similar if their neighbors are similar as well*. Essentially, the similarity of any two objects, g_1 and g_2, can be calculated as follows:

$$s(g_1, g_2) = \frac{DC}{|P_{g_1}||P_{g_2}|} \sum_{i=1}^{|P_{g_1}|} \sum_{j=1}^{|P_{g_2}|} s(P_{g_1}^i, P_{g_2}^j) \tag{9}$$

where DC is a decay factor and $DC \in [0,1]$, P_{g_1} and P_{g_2} are the sets of neighbors of objects g_1 and g_2, respectively. Individual neighbors of these objects are specified as $P_{g_1}^i$ and $P_{g_2}^j$, for $1 \leq i \leq |P_{g_1}|$ and $1 \leq j \leq |P_{g_2}|$. Note that $s(g_1, g_2) = 0$ when $P_{g_1} = \emptyset$ or $P_{g_2} = \emptyset$. It is suggested by Jeh and Widom [19] that the optimal similarity measures could be obtained through iterative refinement of similarity values to a fixed-point (i.e. after k iterations).

$$\lim_{k \to \infty} R_k(g_1, g_2) = s(g_1, g_2) \tag{10}$$

$$R_{k+1}(g_1, g_2) = \frac{DC}{|P_{g_1}||P_{g_2}|} \sum_{i=1}^{|P_{g_1}|} \sum_{j=1}^{|P_{g_2}|} R_k(P_{g_1}^i, P_{g_2}^j) \tag{11}$$

At the outset, this iterative process starts off using the lower bound of: $R_0(g_1, g_2) = 1$ if $g_1 = g_2$, and 0 otherwise.

Applying SimRank to the Cluster Ensemble Problem. Besides considering a cluster ensemble as a network of clusters only (as with the CTS method), a bipartite representation can be utilized to reveal additional hidden relations. Figure 4(a) and 4(b) show the cluster results of two base clusterings, and the corresponding bipartite graph is presented in Figure 4(c).

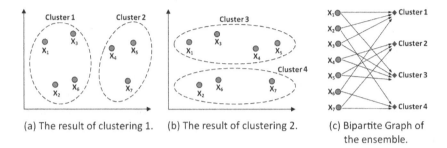

(a) The result of clustering 1.   (b) The result of clustering 2.   (c) Bipartite Graph of
                                                                       the ensemble.

Fig. 4. Representing a cluster ensemble as a bipartite graph

Given a cluster ensemble Π, a graph $G = (V, E)$ can be constructed, where V is a set of vertices representing both data points and clusters in the ensemble and E denotes a set of edges between data points and their clusters. Let $SRS(a, b)$ be an entry in the SRS matrix, which represents the similarity between any pair of data points or the similarity between any two clusters in the ensemble. For $a = b, SRS(a, b) = 1$. Otherwise,

$$SRS(a, b) = \frac{DC}{|N_a||N_b|} \sum_{a' \in N_a} \sum_{b' \in N_b} SRS(a', b') \tag{12}$$

where DC is constant decay factor within the interval $[0, 1]$, N_x denotes the set of vertices connecting to x. As a result, the similarity between data points x_i and x_j is the average similarity between the clusters to which they belong, and the similarity between clusters is the average similarity between their members.

Iterative Computation of SimRank. The similarity scores between any pair of vertices can be computed through the iteration process. Let $SRS_r(a, b)$ be a similarity score between a and b at iteration r, the estimation of the similarity score at the next iteration $r + 1$ is given as

$$SRS_{r+1}(a, b) = \frac{DC}{|N_a||N_b|} \sum_{a' \in N_a} \sum_{b' \in N_b} SRS_r(a', b') \tag{13}$$

Note that, at the outset, $SRS_0(a, b) = 1$ if $a = b$ and 0 otherwise.

4   Complexity Improvements via Approximate Methodologies

The major drawback of pairwise similarity methods is the high computational complexity, especially with the link-based matrices, which limits their application only to small and medium-size data [13], [16]. Given N data points and M ensemble members (i.e. base clusterings), the time complexity of creating the CO matrix is $O(N^2 M)$, while that of the CTS matrix is $O(N^2 MT_1)$ (where T_1 is the average of $|N_{C_a}||N_{C_b}|$, C_a and C_b are the clusters to which data points a and b belong, $|N_{C_a}|$ and $|N_{C_b}|$ are the number of clusters that share members with C_a and C_b, respectively). In addition, the same requirement of the SRS matrix is $O(r(N^2 T_2 + C^2 T_3))$, where T_2 is the average of $|N_a||N_b|$ over all pairs of data points (a, b), N_a and N_b are the set of clusters linked to data points a and b, respectively. Similarly, T_3 denotes the average of $|N_c||N_d|$ over all pairs of clusters (c, d), N_c and N_d are the set of data points linked to clusters a and b. With the SimRank algorithm, r is the number of iterations of estimating similarity values and C is the total number of clusters in the ensemble.

As a result, computing the two link-based matrices is more computationally expensive than computing the original CO matrix, and their use may be impractical for large datasets. However, as empirically demonstrated in [16], they greatly improve robustness and quality of clustering results, by being able to recover additional hidden relations among data points that are completely neglected in the CO approach. Therefore, this section introduces two methodologies for reducing computational complexity of link-based matrices and extending their applicability to large datasets: (i) approximating the similarity estimation performance of the SRS matrix with the new ASRS (Approximate SRS) approach and (ii) approximating the data via a prototyping model that can preserve the quality of similarity matrices found from the original data.

4.1   Approximate SimRank Based Similarity (ASRS) Method

With the purpose of enhancing the applicability of the SRS approach, the ASRS method is introduced to reduce the computational requirement by eliminating the iterative SimRank process. Essentially, this new similarity matrix is proposed upon the same assumption used in the SRS counterpart that the similarity between data points a and b is the average similarity between the clusters to which they belong.

At the outset, a bipartite graph $G = (V, E)$ is constructed to represent a cluster ensemble Π, where V is a set of vertices representing both data points and clusters in the ensemble and E denotes a set of edges between data points and their clusters. Let $ASRS(a, b)$ is the entry in the ASRS matrix, which represents the similarity between any pair of data points in the ensemble. For $a = b, ASRS(a, b) = 1$. Otherwise,

$$ASRS(a, b) = \frac{1}{|N_a||N_b|} \sum_{a' \in N_a} \sum_{b' \in N_b} wSRS(a', b') \qquad (14)$$

where N_x denotes the set of vertices connecting to data point x (i.e. a set of clusters to which x belongs) and $wSRS(y, z)$ is a similarity value between clusters y and z, which can be obtained using the weighted SimRank algorithm described below.

Weighted SimRank. Given a cluster ensemble Π, a graph $G = (V, W)$ can be constructed where V is the set of vertices each representing a cluster in Π and W is a set of weighted edges between clusters. Formally, the weight assigned to the edge connecting clusters i and j is estimated in accordance with the proportion of their overlapping members.

$$w_{ij} = \frac{|X_i \cap X_j|}{|X_i \cup X_j|} \tag{15}$$

where X_A denotes the set of data points belonging to cluster A. Let $wSRS(y, z)$ be a similarity between any two clusters. For $y = z$, $wSRS(y, z) = 1$. Otherwise, it can be estimated as follows.

$$wSRS(y, z) = \frac{wSR(y, z)}{wSR_{max}} DC \tag{16}$$

where $DC \in [0, 1]$ is the confidence level to accept two non-identical clusters to be similar, and wSR_{max} is the maximum wSR value of any two clusters y and z, being defined as

$$wSR(y, z) = \frac{1}{|N_y||N_z|} \sum_{y' \in N_y} \sum_{z' \in N_z} (w_{yy'} \times w_{zz'}) \tag{17}$$

where N_y and N_z are the set of clusters to which clusters y and z are linked (i.e. sharing data points), respectively.

Using the prescribed approach of ASRS, the time complexity required for estimating the pairwise similarity amongst data points is reduced from $O(r(N^2 T_2 + C^2 T_3))$ (with the SRS method) to $O(N^2 T_2 + C^2 T_1)$, where T_3 measured in a bi-partite network is typically greater than T_1 estimated in the single-object network of clusters.

4.2   Approximate Data: The Prototype Based Cluster Ensemble Method

Another approximating approach to help pairwise similarity methods scale up to large datasets is the prototype-based cluster ensemble scheme. This new methodology involves three phases: (i) selecting representative data or prototypes, (ii) performing cluster ensemble on the set of prototypes using the pairwise similarity methods and (iii) mapping the original dataset to the clustering solution achieved from the cluster ensemble process of prototypes. The generic process of this approximating model is demonstrated in Figure 5.

Specifically, from a given dataset $X = \{x_1, \ldots, x_N\}$ of N data points, a set of prototypes $R = \{r_1, \ldots, r_P\}$ is firstly generated, where P is the number of prototypes and $P < N$. Subsequently, pairwise similarity methods are

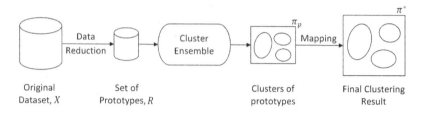

Fig. 5. The process of prototype-based cluster ensemble

applied to the set of prototypes and provide a set of clusters of prototypes $\pi_p = \{C_1^p, C_2^p, \ldots, C_k^p\}$, where k is the desired number of clusters. Finally, each data point in the original dataset X is mapped to the solution of the previous cluster ensemble process π_p, to obtain the final clustering partition π^*.

Initially, selecting a subset of representative data (i.e. prototypes) from the original dataset is performed using data sampling techniques, which have been extensively exploited to reduce computational complexity of data mining algorithms. The most commonly used is the uniform random sampling with which each object has an equal probability of being included in the sample set. However, in the case of datasets with skewed cluster sizes, this simple technique often fails to select data points from small clusters [30]. To overcome this shortcoming, Density-Biased Sampling (DBS) was first introduced in [30], with which the probability of including a data point in the sample set is based on local density of its neighbors. This sampling methodology has proven more effective than uniform sampling [22], [24], [30]. However, most of DBS algorithms are sensitive to noise and some of them are suffered from high computational requirements.

To this extend, another density-biased sampling algorithm, named Biased Box Sampling (BBS) [1], was developed to be less sensitive to noise and able to provide a superior set of representative objects within linear time complexity of $O(NE)$, where N and E are the number of data points in the original dataset and the number of attributes, respectively. Henceforth, the proposed prototype-based model makes use of this BBS sampling technique to generate a set of prototypes.

Having applied a pairwise similarity method to selected prototypes, a mapping mechanism is then exploited to assign each data point to an appropriate cluster of prototypes previously derived. In particular, each original data point $x_i \in X$ is assigned to the cluster of prototypes C_*^p, with the minimum of average pairwise distance between x_i and all prototypes in that cluster.

$$C_*^p = C_j^p, \min_{\forall C_j^p, j=1\ldots k} \frac{\sum_{\forall r_s \in C_j^p, s=1\ldots P_j} d(x_i, r_s)}{P_j} \tag{18}$$

where $C_j^p \in \pi^p, j = 1\ldots k$, P_j is the number of prototypes in the cluster C_j^p, $\sum_{j=1\ldots k} P_j = P$ (i.e. the total number of prototypes), and $d(x_i, r_s)$ denotes the distance between a data point x_i and the prototype r_s.

5   Performance Evaluation

This section evaluates the performance of different pairwise similarity methods, using a variety of validity indices, over both synthetic and real-world datasets. The applicability of these techniques, which is enhanced by the approximation of link-based similarity estimation and a data prototyping technique, is empirically studied in several settings of cluster ensemble.

5.1   Datasets

Four pairwise similarity matrices (CO, CTS, SRS and ASRS) are experimentally evaluated over 23 datasets, where true natural clusters are known but are not explicitly used by the cluster ensemble process. The details of these datasets are summarized in Table 1, which divided into three categories of: small, medium and large.

Table 1. Description of datasets: number of data points, number of features, number of classes and source

Dataset	Data points	Features	Classes	Source
Small datasets:				
Difficult doughnut	100	12	2	[26]
4-gaussian	100	12	4	[26]
2-doughnut	100	3	2	[5]
2-spiral	190	2	2	[5]
2-banana	200	2	2	[16]
Iris	150	4	3	UCI [2]
Wine	178	13	3	UCI [2]
Glass	214	9	6	UCI [2]
Ecoli	336	8	8	UCI [2]
Ionosphere	351	34	2	UCI [2]
Medium datasets:				
Complex Image	500	2	11	modified from [26]
5-gaussian	600	2	5	modified from [26]
3-ring	600	2	3	modified from [26]
Breast Cancer	683	10	2	UCI [2]
Pima Indians Diabetes	768	8	2	UCI [2]
Vehicle	846	18	4	UCI [2]
Large datasets:				
Yeast	1,484	9	10	UCI [2]
Image Segmentation	2,310	19	7	UCI [2]
Optical Digits	3,823	64	10	UCI [2]
Spambase	4,601	57	2	UCI [2]
Landsat Satellite	6,435	36	6	UCI [2]
Pen Digits	10,992	16	10	UCI [2]
Census Income	299,285	7	2	UCI [2]

Eight synthetic datasets are included in the experiments: Difficult doughnut, 4-gaussian, 2-doughnut, 2-spiral, 2-banana, Complex Image, 5-gaussian and
3-ring, shown in Figure 6(a) to 6(h), respectively. Particularly, the first two synthetic datasets acquired from [26] are created in two dimensions with added ten
more dimensions of noise. In addition to the synthetic data collection, 14 real-
world datasets obtained from UCI benchmark repository [2] are also employed.
Specific to Census Income dataset that originally consists of seven continuous
and thirty three nominal features, only those continuous ones are included in
the current experiment of numerical data clustering.

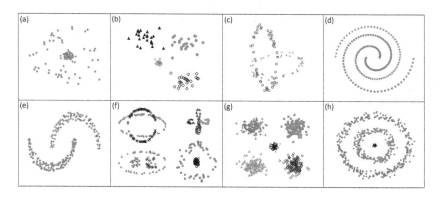

Fig. 6. Synthetic datasets: (a) Difficult doughnut, (b) 4-gaussian, (c) 2-doughnut,
(d) 2-spiral, (e) 2-banana, (f) Complex Image, (g) 5-gaussian and (h) 3-ring

5.2   Evaluation Criteria

Since the external class labels are available for all experimented datasets, the
results of final clustering are evaluated using three label-oriented validity indices:
Classification Accuracy [29], Normalized Mutual Information [32] and Rand
index [31]. The label-oriented validity index assesses the degree of agreement
between two data partitions, where one of the partitions is obtained from a clustering algorithm (π^*) and the other is taken from a prior information, i.e. the
known label of the data (Π'). The description of each validity index is given
below.

Classification Accuracy. The classification accuracy (CA) is commonly used
for evaluating clustering results. It measures the number of correctly classified
data points of a clustering solution compared with known class labels. To compute the CA, each cluster from the clustering result is relabeling with the *majority* class label, which most of data points in that cluster come from, and then
the accuracy of the new labels is measured by counting the number of correctly
labeled data points, in comparison to their known class labels, and dividing by
the total number of data in the dataset.

Let m_i is the number of data points with the *majority* class label in cluster i, the CA can be regarded as the ratio of the number of correctly classified data points to the total number of data points in the dataset. According to the definition given by Nguyen and Caruana [29], the CA is defined as,

$$CA(\pi^*, \Pi') = \frac{\sum_{i=1}^{K}(m_i)}{N} \tag{19}$$

where N is the total number of data in the dataset. The CA ranges from 0 to 1. If the clustering result takes value 1 of the CA, it denotes that all data points are clustered correctly and the clustering contains only pure clusters, i.e. clusters in which all data points have the same class label.

Normalized Mutual Information. This criterion is proposed by Strehl and Ghosh [32] to measure how similar the two data partitions are. The Normalized Mutual Information (NMI) measures the average mutual information between every pair of cluster and class. Its range is $0 < NMI \leq 1$ and the maximum value of 1 indicating that the clustering result and the original classes completely match. Given the two data partitions with K clusters and K' classes, respectively, the NMI is computed by the following equation.

$$NMI(\pi^*, \Pi') = \frac{\sum_{i=1}^{K}\sum_{j=1}^{K'} n_{i,j} \log(\frac{n_{i,j}N}{n_i m_j})}{\sqrt{\sum_{i=1}^{K} n_i \log(\frac{n_i}{N})\sum_{j=1}^{K'} m_j \log(\frac{m_j}{N})}} \tag{20}$$

where $n_{i,j}$ is the number of data points agreed by cluster i and class j, n_i is the number of data points in cluster i, m_j is the number of data points in class j and N is the total number of data points in the dataset.

Rand Index. The Rand index [31] is one of many validity indices that evaluate the agreement between two data partitions. It takes into account the number of object pairs that exist in the same and different clusters. More formally, the Rand index (RI) can be defined as

$$RI(\pi^*, \Pi') = \frac{n_{11} + n_{00}}{n_{11} + n_{10} + n_{01} + n_{00}} \tag{21}$$

where n_{11} is the number of pairs of data points that are in the same clusters in both partitions π^* and Π', n_{00} denotes the number of pairs of data points that are placed the different clusters in both π^* and Π', n_{10} is the number of pairs of data points, which belong to the same cluster in π^* but are in the different clusters in Π', and n_{01} indicates the number of pairs of data points, which are put in the different clusters in π^* but are in the same cluster in Π'.

Intuitively, n_{11} and n_{00} can be interpreted as the quantity of agreements between two partitions, while n_{10} and n_{01} are the number of disagreements. The Rand index has a value between 0 and 1, with the more the value approximates to 1 the higher the agreement is.

5.3   Empirical Evaluation of Approximate SimRank Based Similarity Method

In order to evaluate quality of the four pairwise similarity matrices, they are empirically compared, over 16 small-medium datasets, using several settings of cluster ensembles exhibited below.

- The k-means clustering algorithm is specifically used to generate the base clusterings, with random initialization of cluster centers.
- Two schemes for selecting the number of clusters (k) in each base clustering are: fixed $k = \sqrt{N}$ and random k in $[2, \sqrt{N}]$, where N is the number of data points.
- Three different ensemble sizes of 10, 20 and 30 base clusterings are experimented, respectively.
- The constant decay factor (DC) are set to be 0.5, 0.8 and 0.8 for the Connected-Triple algorithm, SimRank algorithm and Approximate SimRank method, respectively.
- The number of iterations for SimRank algorithm is set to be 4.
- Consensus methods: three agglomerative approaches (single-linkage: SL, complete-linkage: CL, and average-linkage: AL) and a graph partitioning method (METIS). Note that, applying METIS to the CO matrix is the technique named *Cluster-based Similarity Partitioning Algorithm (CSPA)* [32]. For comparison purpose, as in [15] and [9], these consensus functions divide data points into K (the number of *true classes* for each dataset) partitions in accordance with the underlying similarity matrix (CO, CTS, SRS or ASRS).
- The ultimate clustering results are evaluated using three validity indices emphasized in Section 5.2. The quality of each similarity matrix with each specific ensemble setting is generalized as the average of 50 runs.

Tables 2 and 3 present two specific subsets of experimental results (using four different consensus methods, the classification accuracy measure and the ensemble size of 30) with small and medium datasets, respectively. As a result, the performance of ASRS is generally competitive to that of the SRS method. Interestingly, in some cases, ASRS can significantly improve the quality of clustering results, especially with average-linkage being employed as the consensus function. In addition, the ASRS matrix usually outperforms the original CO method, especially with the complete-linkage algorithm where the CO matrix causes the worst performance across all experimented datasets. This empirical evidence effectively implies the better quality of three link-based similarity matrices comparing to the traditional co-association method.

In order to further evaluate the quality of four similarity matrices over each dataset, the number of times that one method is significantly better (of 95% confidence level) than the others are assessed across all experimental settings. Let $\overline{X}_C(i, j)$ be the average value of validity index C across n runs for a similarity matrix $j \in SM$ $(SM = \{CO, CTS, SRS, ASRS\})$ and a specific compositional setting i (from $4 \times 2 \times 3$ different combinations of consensus function, base

Table 2. Classification accuracy (in percentage) of CO, CTS, SRS and ASRS pairwise methods. Represented figures are the averages across 50 runs of cluster ensemble (size = 30), using four different consensus techniques (SL, CL, AL and METIS) and two ensemble generating schemes (Fixed k and Random k), over ten small datasets. The highest CA score of each specific ensemble setting is highlighted in **boldface**.

Dataset	Consensus Function	Fixed k				Random k			
		CO	CTS	SRS	ASRS	CO	CTS	SRS	ASRS
Difficult	SL	62.90	63.82	**93.04**	88.44	90.50	92.40	96.58	**97.62**
doughnut	CL	60.04	71.22	**74.80**	63.50	67.30	76.00	**76.58**	76.52
	AL	93.60	88.02	97.66	**97.80**	72.42	74.88	79.84	**85.62**
	METIS	98.68	**99.00**	**99.00**	98.96	**99.02**	99.00	99.00	99.00
4-gaussian	SL	93.84	94.80	96.16	**98.28**	97.96	98.00	**98.56**	90.94
	CL	77.74	98.52	98.52	**98.68**	95.72	97.76	**98.40**	98.40
	AL	98.68	98.56	98.60	**98.74**	98.20	98.32	**98.42**	98.24
	METIS	99.16	99.26	**99.34**	98.90	**99.34**	98.60	98.72	98.80
2-doughnut	SL	100.00	100.00	100.00	100.00	73.82	72.96	**81.50**	80.34
	CL	63.28	73.64	**87.80**	78.78	67.74	**72.44**	71.80	69.96
	AL	92.94	98.22	**99.22**	96.48	66.02	66.34	67.06	**67.66**
	METIS	90.16	99.02	100.00	95.16	61.16	67.98	68.36	**71.76**
2-spiral	SL	**61.62**	59.01	57.49	51.21	53.94	53.59	**61.61**	51.29
	CL	52.46	**55.93**	50.69	55.43	57.20	**61.37**	60.08	61.34
	AL	53.16	52.41	**54.13**	53.93	61.85	62.12	**62.47**	62.06
	METIS	59.36	60.73	**63.49**	60.89	64.06	**65.00**	64.47	64.85
2-banana	SL	100.00	100.00	100.00	100.00	100.00	100.00	100.00	100.00
	CL	58.06	66.18	**75.09**	71.76	70.55	87.59	**88.78**	88.07
	AL	100.00	100.00	100.00	100.00	95.07	97.92	97.65	**99.81**
	METIS	100.00	100.00	100.00	100.00	86.62	94.14	92.93	**96.94**
Iris	SL	76.60	79.09	83.09	**92.64**	81.80	81.31	**90.43**	84.00
	CL	52.57	83.93	**85.08**	84.53	82.89	**88.47**	87.71	87.45
	AL	86.81	83.89	87.25	**92.56**	85.45	85.85	85.73	**87.53**
	METIS	95.64	96.04	96.08	**96.17**	95.44	95.59	95.48	**95.72**
Wine	SL	57.53	58.14	60.06	**75.35**	51.42	53.11	**67.73**	60.58
	CL	50.11	91.47	**92.34**	92.33	79.75	**92.87**	92.04	91.92
	AL	89.00	83.47	91.06	**91.98**	**95.83**	95.64	94.99	95.44
	METIS	92.16	92.30	92.06	**92.33**	92.82	92.69	92.71	**93.28**
Glass	SL	51.18	51.26	**59.33**	51.00	46.12	46.19	**50.55**	48.64
	CL	49.93	53.94	**54.49**	53.41	51.96	52.35	52.17	**52.95**
	AL	51.25	51.10	52.40	**52.47**	**53.00**	52.31	52.68	52.44
	METIS	**61.18**	59.99	59.83	60.69	58.81	59.58	59.97	**60.67**
Ecoli	SL	71.70	**72.36**	51.78	63.05	74.10	**75.46**	75.14	72.11
	CL	56.76	80.81	**80.98**	79.39	79.89	**80.43**	79.83	79.79
	AL	**83.86**	83.77	83.61	81.40	81.32	**81.74**	81.38	80.13
	METIS	**77.76**	77.38	76.36	76.70	**77.26**	76.82	76.43	77.04
Ionosphere	SL	65.30	65.47	64.82	**65.97**	64.90	65.15	**65.64**	64.52
	CL	67.15	**77.46**	66.45	66.84	**74.38**	70.73	71.20	68.87
	AL	70.85	72.76	**73.02**	67.98	71.34	71.30	72.78	**73.18**
	METIS	**67.30**	66.29	67.05	66.34	68.48	**68.50**	68.49	67.59

clustering generation scheme and ensemble size). The 95% confidence interval for the mean $\overline{X}_C$ of each validity criterion C is calculated as

$$\left[\overline{X}_C(i,j) - 1.96\frac{S_C(i,j)}{\sqrt{n}}, \overline{X}_C(i,j) + 1.96\frac{S_C(i,j)}{\sqrt{n}}\right] \quad (22)$$

where $S_C(i,j)$ denotes the standard deviation of the validity index C across n runs for experiment setting i and similarity matrix j.

The statistical significance of the difference between any two similarity matrices, p and q, on any experiment setting, i, is observed if there is no intersection

Table 3. Classification accuracy (in percentage) for the four similarity matrices over six medium datasets

Dataset	Consensus Function	Fixed k				Random k			
		CO	CTS	SRS	ASRS	CO	CTS	SRS	ASRS
Complex Image	SL	70.09	70.26	**73.71**	71.67	**71.12**	69.22	67.86	49.79
	CL	53.05	58.52	57.64	**60.03**	55.74	**56.46**	54.25	52.38
	AL	62.02	63.12	63.65	**64.47**	**60.93**	60.78	59.03	53.38
	METIS	**65.81**	65.63	65.56	65.67	64.73	64.66	64.75	**65.78**
5-gaussian	SL	100.00	100.00	100.00	100.00	100.00	100.00	100.00	100.00
	CL	43.71	100.00	100.00	100.00	99.79	100.00	100.00	100.00
	AL	100.00	100.00	100.00	100.00	100.00	100.00	100.00	100.00
	METIS	100.00	100.00	100.00	100.00	100.00	100.00	100.00	100.00
3-ring	SL	100.00	100.00	100.00	100.00	**87.29**	86.64	83.51	76.70
	CL	50.00	54.18	**63.17**	55.66	**52.78**	51.55	51.46	51.03
	AL	100.00	100.00	100.00	100.00	51.97	53.58	52.73	**55.17**
	METIS	82.43	83.05	**83.39**	83.08	59.22	58.75	59.43	**66.77**
Breast Cancer	SL	65.64	65.57	66.39	**69.36**	79.34	**82.62**	73.43	66.94
	CL	66.51	80.43	**96.04**	77.24	76.97	**96.98**	96.57	96.75
	AL	**96.78**	96.72	96.28	96.04	96.06	96.97	**96.98**	96.90
	METIS	75.04	**85.07**	85.06	85.01	83.20	84.59	84.77	**84.81**
Pima Indians Diabetes	SL	**65.63**	65.56	65.56	65.37	65.10	65.10	65.12	**65.18**
	CL	65.10	**65.84**	65.44	65.55	65.22	65.61	**66.20**	66.03
	AL	65.10	65.10	65.10	65.10	**65.51**	65.39	65.41	65.38
	METIS	65.10	**65.23**	65.17	65.12	**67.38**	65.71	66.62	67.08
Vehicle	SL	**30.90**	30.76	29.44	27.32	29.64	**29.84**	28.41	26.34
	CL	29.48	37.36	**40.16**	39.60	34.44	38.63	**38.93**	38.80
	AL	39.77	**40.16**	39.75	39.54	**40.83**	40.82	40.76	40.31
	METIS	40.92	45.34	45.22	**45.44**	41.66	**42.80**	41.54	42.28

between their confidence interval. Formally, for any setting i, a similarity matrix p is significantly better than another matrix q when

$$\left(\overline{X}_C(i,p) - 1.96 \frac{S_C(i,p)}{\sqrt{n}} \right) > \left(\overline{X}_C(i,q) + 1.96 \frac{S_C(i,q)}{\sqrt{n}} \right) \tag{23}$$

For each dataset, Table 4 presents the number of times that one similarity matrix $p \in SM$ is significantly better than its competitors, $better\_score_C(p)$, in accordance with the validity criterion $C \in \{CA, NMI, RI\}$, across all experimental settings.

$$better\_score_C(p) = \sum_{\forall i} \sum_{\forall q \in SM, q \neq p} better_C^i(p,q), \forall p \in SM \tag{24}$$

$$better_C^i(p,q) = \begin{cases} 1 \ \ if \ \left(\overline{X}_C(i,p) - 1.96 \frac{S_C(i,p)}{\sqrt{n}} \right) > \left(\overline{X}_C(i,q) + 1.96 \frac{S_C(i,q)}{\sqrt{n}} \right) \\ 0 \ \ otherwise \end{cases} \tag{25}$$

where i denotes a particular setting of cluster ensemble (i.e. a specific combination of consensus function, base clustering generation scheme and ensemble size). According to this statistics, the three link-based approaches generally perform much better than the conventional CO method across all validity measures.

Table 4. The number of times that each matrix provides the better performance (at 95% confidence level), measured by three validity indices, across 16 datasets, four combination methods (SL, CL, AL and METIS), two different ensemble distributions (Fixed $k = \sqrt{N}$ and Random k in $[2, \sqrt{N}]$) and three ensemble sizes (10, 20 and 30)

Dataset	CA				NMI				RI			
	CO	CTS	SRS	ASRS	CO	CTS	SRS	ASRS	CO	CTS	SRS	ASRS
Difficult doughnut	0	9	29	30	0	8	26	31	0	9	29	30
4-gaussian	6	7	12	8	5	9	12	7	6	9	12	8
2-doughnut	0	10	17	10	1	10	17	10	0	10	17	8
2-spiral	5	13	23	9	3	4	20	3	3	10	25	9
2-banana	0	8	13	14	0	8	13	14	0	8	13	14
Iris	0	8	14	30	0	8	13	34	0	8	13	32
Wine	1	8	15	27	1	8	14	24	1	8	14	26
Glass	2	3	23	11	5	11	7	7	0	4	30	11
Ecoli	15	19	14	5	11	36	12	6	12	33	13	4
Ionosphere	14	14	8	20	15	19	12	7	17	13	9	23
Complex Image	12	13	17	12	13	13	13	16	11	12	17	8
5-gaussian	0	5	5	5	0	5	5	5	0	5	5	5
3-ring	1	4	10	11	2	5	9	12	1	7	13	14
Breast Cancer	4	22	21	19	3	22	25	19	4	22	22	19
Pima Indians Diabetes	6	8	11	12	10	11	17	21	7	9	16	18
Vehicle	7	17	17	10	9	20	14	10	4	16	18	10
Total	73	168	249	233	78	197	229	226	66	183	266	239

In addition, Figure 7 presents a bar chart which compares the frequency of better performance, detailed by four consensus functions, between the four matrices across all experiment settings and 16 datasets. The results indicate that the CO matrix performs worst over all consensus methods, compared to the others, especially with the CL algorithm. Amongst the three link-based methods, the SRS matrix greatly outperforms the other two in the CL case. However, with SL and METIS, it provides comparably results with the ASRS matrix. More interestingly, with the AL consensus function, ASRS can notably enhance the performance of the SRS counterpart. Hence, this bar graph signifies that the ASRS method can produce the clustering results with comparative quality to the SRS approach, however, with less time complexity.

5.4 Empirical Evaluation of Prototype Based Cluster Ensemble Approach

While the previous section demonstrated the performance of the approximate algorithm, Approximate SimRank Based Similarity (ASRS) matrix, this section evaluates the quality of the other approximate technique, the prototype-based cluster ensemble approach. Two sets of experiments are conducted over: (i) four small-size datasets (Difficult doughnut, 4-gaussian, Iris and Wine) and (ii) seven large datasets containing $1,000$ to $300,000$ instances. Firstly, with the experimental setting described in Section 5.3, the quality of clustering results obtained from the prototype model are compared to those acquired by using original datasets. For the prototype approach, the methods of selecting the number of clusters (k) in each base clustering (k-means) are modified to be fixed $k = \sqrt{P}$ and random $k \in [2, \sqrt{P}]$, where P is the number of prototypes. Additionally,

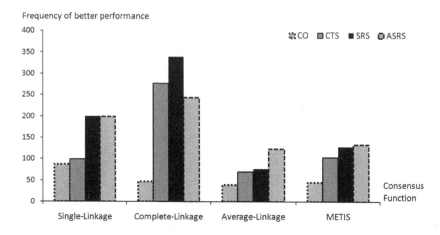

Fig. 7. The number of time that each pairwise similarity matrix provides significantly better performance (at 95% confidence level) than other counterparts, measured by three different validity criteria, over 16 small-medium experimented datasets and categorized by four consensus functions

parameter values of the BBS data sampling algorithm are $R = 3$, $Ratio = 70$, providing the number of prototypes: 31 for the two synthetic datasets, 24 and 96 for Iris and Wine datasets, respectively.

Empirical results achieved using each pairwise similarity method over these four small datasets with ensemble size of 30, shown in Table 5, can be discussed in two folds as follows:

- **Performance of the prototype-based approach against the exploitation of original dataset.** In case of three agglomerative hierarchical consensus functions (SL, CL and AL), the prototype-based approach frequently improves the quality of clustering results, especially over 4-gaussian and two real datasets. Particularly to the 4-gaussian dataset, the prototype model usually provides better performance than the original-data counterpart for all similarity matrices with the SL, and rather competitive with the CL and AL algorithms.

  In contrary, using METIS as the consensus function results in worse performance than that of using the original data, especially over the first three balanced datasets (i.e. Difficult doughnut, 4-gaussian and Iris). This is due to the fact that generating a set of prototypes may no longer promise balanced groups of prototypes, which METIS attempts to partition into equally-size clusters. As a result, assigning original data points to wrong groups of prototypes leads to low quality of final clustering results. However, with an unbalanced dataset like Wine, results obtained from METIS are improved with all similarity matrices.

- **Performance of four similarity matrices in prototype approach.** For Difficult doughnut dataset, link-based matrices usually outperforms the CO

Table 5. Classification accuracy (in percentage) averaged across 50 runs for each matrix over four small datasets, two ensemble generating schemes (Fixed k and Random k), ensemble size of 30, four consensus functions (SL, CL, AL and METIS), with both original data (Org) and prototype paradigm (Ptt)

Dataset	Base Clustering	Consensus Function	CO Org	CO Ptt	CTS Org	CTS Ptt	SRS Org	SRS Ptt	ASRS Org	ASRS Ptt
Difficult doughnut	Fixed k	SL	62.90	67.86	63.82	68.00	93.04	68.00	88.44	68.00
		CL	60.04	62.48	71.22	70.18	74.80	69.08	63.50	69.92
		AL	93.60	68.06	88.02	68.00	97.66	68.00	97.80	68.00
		METIS	98.68	65.02	99.00	66.76	99.00	66.54	98.96	66.08
	Random k	SL	90.50	70.88	92.40	70.90	96.58	68.14	97.62	69.38
		CL	67.30	67.18	76.00	70.98	76.58	71.16	76.52	70.92
		AL	72.42	70.84	74.88	71.00	79.84	70.76	85.62	70.74
		METIS	99.02	65.82	99.00	65.84	99.00	65.78	99.00	65.80
4-gaussian	Fixed k	SL	93.84	98.00	94.80	98.00	96.16	98.00	98.28	98.00
		CL	77.74	98.00	98.52	98.00	98.52	98.00	98.68	98.00
		AL	98.68	98.00	98.56	98.00	98.60	98.00	98.74	98.00
		METIS	99.16	87.62	99.26	87.30	99.34	87.66	98.90	89.02
	Random k	SL	97.96	98.00	98.00	98.00	98.56	98.00	90.94	98.00
		CL	95.72	98.00	97.76	98.00	98.40	98.00	98.40	98.00
		AL	98.20	98.00	98.32	98.00	98.42	98.00	98.24	98.00
		METIS	99.34	88.14	98.60	88.14	98.72	87.94	98.80	88.12
Iris	Fixed k	SL	76.60	84.67	79.09	84.67	83.09	84.67	92.64	84.67
		CL	52.57	84.67	83.93	84.67	85.08	84.67	84.53	84.67
		AL	86.81	84.67	83.89	84.67	87.25	84.67	92.56	84.67
		METIS	95.64	84.83	96.04	85.11	96.08	85.16	96.17	84.47
	Random k	SL	81.80	84.67	81.31	84.67	90.43	84.67	84.00	84.67
		CL	82.89	84.67	88.47	84.67	87.71	84.67	87.45	84.67
		AL	85.45	84.67	85.85	84.67	85.73	84.67	87.53	84.67
		METIS	95.44	85.09	95.59	85.11	95.48	84.69	95.72	85.35
Wine	Fixed k	SL	57.53	90.39	58.14	90.62	60.06	89.94	75.35	92.30
		CL	50.11	67.06	91.47	94.87	92.34	94.85	92.33	94.88
		AL	89.00	94.84	83.47	94.92	91.06	94.89	91.98	94.83
		METIS	92.16	94.90	92.30	94.75	92.06	94.91	92.33	92.83
	Random k	SL	51.42	89.74	53.11	91.80	67.73	92.54	60.58	93.78
		CL	79.75	92.40	92.87	93.99	92.04	94.79	91.92	95.19
		AL	95.83	95.27	95.64	95.21	94.99	95.16	95.44	95.04
		METIS	92.82	94.91	92.69	94.93	92.71	94.94	93.28	94.90

matrix with *Fixed k* scheme, while competitive results are obtained in the case of *Random k* setting. With 4-gaussian dataset, four matrices achieve quite consistent and similar results with all consensus functions, and agglomerative hierarchical algorithms in particular. Interestingly, similar observations can also be found over Iris dataset. As for Wine dataset, the performance of the CO matrix is often worse than the other three.

By following the same experiment settings described in Section 5.3, the second set of experiments evaluates the prototype-based cluster ensemble approach over seven large datasets from UCI repository [2], as described in Table 1, using the same base clustering generating schemes as the previous set of experiments. However, there are further adjustments regarding parameters of the BBS algorithm in order to produce prototype sets of around 300 to 400 instances as shown in Table 6. Furthermore, as suggested in the previous experiments, METIS does not perform well with unbalanced prototype sets, only three agglomerative hierarchical consensus functions are thus examined in this empirical study.

Table 6. Description of large datasets: number of data points, number of prototypes and *Ratio* (parameter values for BBS algorithm). Note that parameter R is set to 3 for all datasets and parameter *Ratio* is dataset-specific (as shown in the table).

Dataset	Data points	Prototypes	Ratio
Yeast	1,484	319	20.0
Image Segmentation	2,310	322	3.5
Optical Digits	3,823	345	3.0
Spambase	4,601	356	6.0
Landsat Satellite	6,435	370	3.6
Pen Digits	10,992	383	1.6
Census Income	299,285	398	5.6

Having evaluated over identified large datasets, Table 7 reports the classification accuracy achieved from a fraction of performance evaluation results using only ensembles of size 30. In particular, measures in the third column are acquired from applying the single run of SL, CL and AL to the original data, and accuracy values obtained from prototype-based approach with four pairwise matrices are averages across 50 runs. According to this table, the prototype method usually outperforms the single run of the SL, CL and AL algorithms, throughout all datasets, except in the Spambase case in which results are rather comparable. Note that the results of single run with the original Census Income dataset that have been marked as 'n/a', can not be obtained by the personal computer used in this experiment - Intel(R) Core(TM)2 CPU 6600 @2.40GHz, 2GB RAM. Substantial improvement can be specifically observed with the pairwise methods using the SL algorithm. Considering the performance of the four similarity matrices in the context of prototype approach, the three link-based matrices usually provide superior results to that of the CO method in the case of CL method. In contrast, with the SL and AL algorithms, the performance of four similarity matrices are equally competitive.

In order to further generalize the performance of the four similarity matrices with prototype-based approach, the statistical significant difference of performance between any pair of matrices, previously used in Section 5.3, is observed. Table 8 presents the frequency of better performance by each matrix in three categories of evaluation criteria, detailed in all 7 large datasets. Moreover, the bar graph shown in Figure 8 illustrates that the CO matrix performs the worst for all consensus methods. Once again, this empirical evidence further suggests that link-based pairwise similarity models can produce better clustering solutions than the co-association technique. This also supports the conclusion that the ASRS method can produce the clustering results with comparative quality to the SRS approach.

Table 7. Clustering performance, measured by classification accuracy (in percentage), obtained from the single run of the SL, CL and AL algorithms with the original data (the third column) and those acquired from the prototype-base approach using the four pairwise matrices (CO, CTS , SRS and ASRS), over seven large datasets, two ensemble generating schemes (Fixed $k = \sqrt{P}$ and Random k in $[2, \sqrt{P}]$) and ensemble size of 30. The **bold-faced** values are the best performance (highest CA value) of each specific experiment setting.

Dataset	Consensus Function	Original Data + Single run	Prototype + Pairwise Combination Methods							
			Fixed k				Random k			
			CO	CTS	SRS	ASRS	CO	CTS	SRS	ASRS
Yeast	SL	32.35	**43.01**	43.00	41.90	41.71	43.88	**44.28**	43.68	41.28
	CL	44.27	44.43	50.28	50.46	**50.53**	48.95	50.23	50.66	**51.54**
	AL	33.02	52.52	52.51	52.80	**53.17**	**51.85**	51.35	51.84	50.80
Image	SL	14.76	58.38	59.64	60.29	**62.87**	60.08	61.42	**62.18**	54.55
Segmentation	CL	56.02	61.00	67.87	**68.26**	67.94	65.26	66.61	67.16	**67.49**
	AL	43.59	69.16	**69.26**	69.08	68.94	67.51	**67.61**	67.25	66.63
Optical Digits	SL	10.46	67.45	67.39	**67.83**	62.03	66.63	66.73	**68.02**	59.56
	CL	67.54	61.89	79.91	**81.18**	80.93	77.01	77.92	**78.44**	77.38
	AL	61.76	81.53	81.61	83.78	**85.32**	**78.36**	77.95	78.25	75.11
Spambase	SL	60.62	**60.60**	60.60	60.60	60.60	**60.60**	60.60	60.60	60.60
	CL	60.60	**64.15**	63.99	60.60	60.94	**61.83**	60.60	60.86	60.90
	AL	60.62	**60.60**	60.60	60.60	60.60	**60.60**	60.60	60.60	60.60
Landsat	SL	23.90	68.06	68.69	**69.57**	66.77	69.27	69.06	**69.35**	67.57
Satellite	CL	50.69	66.54	**68.97**	68.97	67.33	68.97	69.16	**69.36**	68.72
	AL	53.49	**68.93**	68.45	68.82	68.89	**68.89**	68.31	68.82	68.27
Pen Digits	SL	10.50	74.30	74.90	75.10	**75.26**	72.98	**73.18**	73.02	64.56
	CL	55.00	66.66	75.01	75.01	**76.10**	71.98	**72.11**	71.89	69.72
	AL	55.59	75.28	76.04	**76.18**	75.93	**74.18**	73.85	73.76	68.46
Census Income	SL	n/a	**93.95**	**93.95**	**93.95**	**93.95**	93.83	93.83	93.83	**93.90**
	CL	n/a	93.80	93.80	**93.83**	**93.83**	93.78	93.78	**93.83**	**93.83**
	AL	n/a	**93.95**	**93.95**	**93.95**	**93.95**	**93.83**	**93.83**	**93.83**	**93.83**

Table 8. The number of times that each matrix provides the significantly better performance, measured by three validity indices, across seven large datasets, three combination methods (SL, CL and AL), two different ensemble distributions (Fixed $k = \sqrt{P}$ and Random k in $[2, \sqrt{P}]$) and three ensemble sizes (10, 20 and 30)

Dataset	CA				NMI				RI			
	CO	CTS	SRS	ASRS	CO	CTS	SRS	ASRS	CO	CTS	SRS	ASRS
Yeast	2	5	6	12	8	17	8	9	0	3	21	27
Image Segmentation	2	6	9	8	5	10	10	3	0	3	16	16
Optical Digits	5	7	9	12	5	8	11	12	4	6	10	11
Spambase	5	3	0	1	4	3	4	6	7	10	0	0
Landsat Satellite	1	3	9	1	3	9	9	3	0	3	7	3
Pen Digits	8	14	11	6	8	15	12	7	5	10	8	11
Census Income	0	0	2	5	0	0	3	4	0	0	2	6
Total	23	38	46	45	33	62	57	44	16	35	64	74

6 Decision-Support Matrix for Alternatives of Pairwise Similarity Method and Approximating Scheme

In this section, based on experimental results previously exhibited, a decision-support matrix is introduced as a tool to help users choosing a pairwise method appropriate for given time availability, data-size and accuracy constraints.

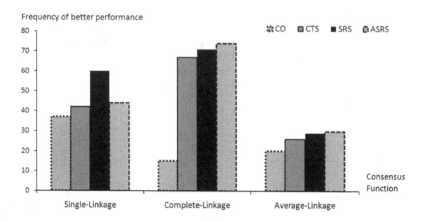

Fig. 8. Bar chart comparing the number of times that each similarity matrix achieves the significantly better clustering results across all experiment settings, measured by three different validity criteria, on each consensus function

Initially, the actual computational time of each method in specific setting is observed. Table 9 lists the average running times (in seconds) for each matrix across 50 repeated run using ensemble size of 10. All matrices were implemented using MATLAB and their execution time have been measured using the cputime() MATLAB's function, on a workstation (Intel(R) Core(TM)2 CPU 6600 @2.40GHz, 2GB RAM). Consistent with the analytic complexity discussed in Section 4, the execution time required by the CO method is the lowest. Considering the three link-based approaches, the CTS approach is up to 2 times faster than the ASRS method, while the SRS is clearly the most time consuming approach. In addition, the cost of computing the ASRS matrix is far less than that of computing the SRS matrix. In fact, the SRS requires 4 to 6 times more execution time than the ASRS method. Moreover, due to the fact that the running time for the three link-based depends on the number of clusters in the ensemble, thus, time spent in *Random k* base clustering scheme is subsequently less than that of *Fixed k* scheme.

Following that, a decision-support matrix is illustrated in Figure 9, where X and Y axis denote the size of data and time allowance, respectively. Particularly, data is roughly categorized into three groups of small (< 500 data points), medium ($500 - 1,000$ data points) and large ($> 1,000$ data points), regarding its size. Similarly, time allowance (t) to studying data is subjectively classified into four intervals (in seconds) of very short ($0 < t \leq 1$), short ($1 < t \leq 5$), moderate ($5 < t \leq 10$) and long ($10 < t$).

For any specific combination of time and data size, a pairwise method is suggested for the best possible accuracy. For instance, given a small-size data and long time availability (i.e. the top-left entry in the matrix), the matrix recommends the SRS method as the most appropriate for such context. For a large dataset, the prototype-based model is exploited since applying pairwise methods

Table 9. Average computational time (in seconds), across 50 runs, for constructing each matrix on a workstation (Intel(R) Core(TM)2 CPU 6600 @2.40GHz, 2GB RAM) using two base clustering schemes (Fixed k and Random k) and ensemble size of 10, over various sizes of experimented datasets

No. of data points	Fixed k				Random k			
	CO	CTS	SRS	ASRS	CO	CTS	SRS	ASRS
100	0.005	0.084	0.500	0.138	0.004	0.025	0.213	0.063
150	0.006	0.175	1.200	0.291	0.004	0.069	0.603	0.159
178	0.007	0.219	1.616	0.381	0.007	0.075	0.709	0.181
190	0.007	0.225	1.725	0.403	0.007	0.066	0.728	0.191
200	0.008	0.266	2.031	0.453	0.008	0.088	0.894	0.225
214	0.011	0.275	2.238	0.488	0.013	0.113	1.134	0.275
336	0.023	0.600	5.444	1.072	0.022	0.203	2.494	0.613
351	0.028	0.616	5.725	1.119	0.025	0.213	2.634	0.634
500	0.054	1.141	11.884	2.103	0.051	0.375	5.156	1.203
600	0.088	1.563	16.685	2.922	0.071	0.513	7.438	1.700
683	0.124	2.034	22.247	3.800	0.104	0.647	9.685	2.203
768	0.169	2.425	27.544	4.653	0.127	0.741	12.066	2.713
846	0.183	2.778	32.150	5.403	0.152	1.088	16.141	3.450

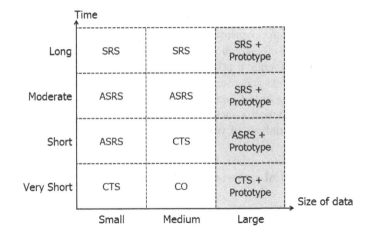

Fig. 9. Decision-support matrix for users' selection of a pairwise method

to original data is greatly expensive. Note that the BBS sampling technique is initially employed to obtain a set of 300 to 400 prototypes, to which the suggested pairwise method is later applied.

7 Conclusion

This paper presents two novel approximate approaches to pairwise similarity methodology for the problem of cluster ensembles: an approximate algorithm approach (Approximate SimRank Based Similarity (ASRS) matrix) and an approximate data approach (Prototype-based cluster ensemble model). The approximate algorithm approach aims to reduce the computational time of the original SimRank Based Similarity (SRS) matrix by eliminating the iterative process of the underlying SimRank algorithm. The empirical studies, with several ensemble settings and validity measures over real-world and synthetic datasets, suggest that the proposed method can provide the comparative performance with the existing SRS approach, however, with substantially less time requirements. In addition, it usually achieves superior clustering results comparing to the traditional co-association (CO) approach.

In spite of this significant enhancement, link-based similarity methods, including ASRS, are less efficient for large-size datasets. Due to this, the second approximate approach–the approximate data approach—sets out to improve this scalability aspect via the application of pairwise similarity techniques to a high-quality set of representative data points (i.e. prototypes), acquired from a density-biased sampling technique. Based on empirical evaluation with several large real-world datasets, the quality of clustering results obtained with the prototype-based method is generally better than those acquired from the single run of hierarchical algorithms over the full original data. Whilst link-based methods are more effective than the CO counterpart, ASRS delivers the performance that is essentially as good as SRS. Finally, a decision-support matrix was provided to aid the appropriate selection of pairwise similarity method and approximate approaches, for given constraints of time, size and accuracy.

Beyond these achievements, it is important to extend the approximate methodology to very large datasets (i.e. more than 1,000,000 instances). We are currently undertaking work in this area to employ our link-based and approximate mechanisms to application domains such as biological data and image analysis that frequently require the analysis of very large datasets.

Acknowledgments. The authors would like to thank Leandro Nunes de Castro for 2-doughnut and 2-spiral datasets, Ana Paula for the executable code of Biased Box Sampling (BSS).

References

1. Appel, A.P., Paterlini, A.A., de Sousa, E.P.M., Traina, A.J.M., Traina Jr., C.: A density-biased sampling technique to improve cluster representativeness. In: Kok, J.N., Koronacki, J., Lopez de Mantaras, R., Matwin, S., Mladenič, D., Skowron, A. (eds.) PKDD 2007. LNCS (LNAI), vol. 4702, pp. 366–373. Springer, Heidelberg (2007)
2. Asuncion, A., Newman, D.J.: UCI machine learning repository (2007)
3. Boulis, C., Ostendorf, M.: Combining multiple clustering systems. In: Boulicaut, J.-F., Esposito, F., Giannotti, F., Pedreschi, D. (eds.) PKDD 2004. LNCS (LNAI), vol. 3202, pp. 63–74. Springer, Heidelberg (2004)
4. Calado, P., Cristo, M., Gonçalves, M.A., de Moura, E.S., Ribeiro-Neto, B.A., Ziviani, N.: Link-based similarity measures for the classification of web documents. JASIST 57(2), 208–221 (2006)
5. de Castro, L.N.: Immune Engineering: Development of Computational Tools Inspired by the Artificial Immune Systems. Ph.D. thesis, DCA - FEEC/UNICAMP, Campinas/SP, Brazil (2001)

6. Domeniconi, C., Al-Razgan, M.: Weighted cluster ensembles: Methods and analysis. ACM Transactions on Knowledge Discovery from Data 2(4), 1–40 (2009)
7. Duda, R.O., Hart, P.E., Stork, D.G.: Pattern Classification, 2nd edn. Wiley-Interscience (November 2000)
8. Fern, X.Z., Brodley, C.E.: Random projection for high dimensional data clustering: A cluster ensemble approach. In: Proceedings of International Conference on Machine Learning, pp. 186–193 (2003)
9. Fern, X.Z., Brodley, C.E.: Solving cluster ensemble problems by bipartite graph partitioning. In: Proceedings of International Conference on Machine Learning, pp. 36–43 (2004)
10. Fred, A.: Finding consistent clusters in data partitions. In: Kittler, J., Roli, F. (eds.) MCS 2001. LNCS, vol. 2096, pp. 309–318. Springer, Heidelberg (2001)
11. Fred, A.L.N., Jain, A.K.: Data clustering using evidence accumulation. In: International Conference on Pattern Recognition, pp. 276–280 (2002)
12. Fred, A.L.N., Jain, A.K.: Robust data clustering. In: International Conference on Pattern Recognition, pp. 128–136 (2003)
13. Fred, A.L.N., Jain, A.K.: Combining multiple clusterings using evidence accumulation. IEEE Transactions on Pattern Analysis and Machine Intelligence 27(6), 835–850 (2005)
14. Fred, A.L.N., Jain, A.K.: Learning pairwise similarity for data clustering. In: International Conference on Pattern Recognition, pp. 925–928 (2006)
15. Gionis, A., Mannila, H., Tsaparas, P.: Clustering aggregation. In: Proceedings of International Conference on Data Engineering, pp. 341–352 (2005)
16. Iam-on, N., Boongoen, T., Garrett, S.: Refining pairwise similarity matrix for cluster ensemble problem with cluster relations. In: Boulicaut, J.-F., Berthold, M.R., Horváth, T. (eds.) DS 2008. LNCS (LNAI), vol. 5255, pp. 222–233. Springer, Heidelberg (2008)
17. Jain, A.K., Law, M.H.C.: Data clustering: A user's dilemma. In: Pal, S.K., Bandyopadhyay, S., Biswas, S. (eds.) PReMI 2005. LNCS, vol. 3776, pp. 1–10. Springer, Heidelberg (2005)
18. Jain, A.K., Murty, M.N., Flynn, P.J.: Data clustering: A review. ACM Computing Survey 31(3), 264–323 (1999)
19. Jeh, G., Widom, J.: Simrank: A measure of structural-context similarity. In: Proceedings of ACM SIGKDD International Conference on Knowledge Discovery and Data Mining, pp. 538–543 (2002)
20. Karypis, G., Aggarwal, R., Kumar, V., Shekhar, S.: Multilevel hypergraph partitioning: applications in VLSI domain. IEEE Transactions on VLSI Systems 7(1), 69–79 (1999)
21. Karypis, G., Kumar, V.: Multilevel k-way partitioning scheme for irregular graphs. Journal of Parallel Distributed Computing 48(1), 96–129 (1998)
22. Kerdprasop, K., Kerdprasop, N., Sattayatham, P.: Density-biased clustering based on reservoir sampling. In: Proceedings of DEXA Workshops, pp. 1122–1126 (2005)
23. Klink, S., Reuther, P., Weber, A., Walter, B., Ley, M.: Analysing social networks within bibliographical data. In: Bressan, S., Küng, J., Wagner, R. (eds.) DEXA 2006. LNCS, vol. 4080, pp. 234–243. Springer, Heidelberg (2006)
24. Kollios, G., Gunopulos, D., Koudas, N., Berchtold, S.: Efficient biased sampling for approximate clustering and outlier detection in large data sets. IEEE Transactions on Knowledge and Data Engineering 15(5), 1170–1187 (2003)
25. Kuncheva, L.I., Hadjitodorov, S.T.: Using diversity in cluster ensembles. In: Proceedings of the IEEE International Conference on Systems, Man and Cybernetics, pp. 1214–1219 (2004)

26. Kuncheva, L.I., Vetrov, D.: Evaluation of stability of k-means cluster ensembles with respect to random initialization. IEEE Transactions on Pattern Analysis and Machine Intelligence 28(11), 1798–1808 (2006)

27. Kyrgyzov, I.O., Maitre, H., Campedel, M.: A method of clustering combination applied to satellite image analysis. In: Proceedings of International Conference on Image Analysis and Processing, pp. 81–86 (2007)

28. Monti, S., Tamayo, P., Mesirov, J.P., Golub, T.R.: Consensus clustering: A resampling-based method for class discovery and visualization of gene expression microarray data. Machine Learning 52(1-2), 91–118 (2003)

29. Nguyen, N., Caruana, R.: Consensus clusterings. In: Proceedings of IEEE International Conference on Data Mining, pp. 607–612 (2007)

30. Palmer, C.R., Faloutsos, C.: Density biased sampling: an improved method for data mining and clustering. SIGMOD Records 29(2), 82–92 (2000)

31. Rand, W.M.: Objective criteria for the evaluation of clustering methods. Journal of the American Statistical Association 66, 846–850 (1971)

32. Strehl, A., Ghosh, J.: Cluster ensembles - a knowledge reuse framework for combining multiple partitions. Journal of Machine Learning Research 3, 583–617 (2002)

33. Swift, S., Tucker, A., Vinciotti, V., Martin, N., Orengo, C., Liu, X., Kellam, P.: Consensus clustering and functional interpretation of gene-expression data. Genome Biology 5, R94 (2004)

34. Topchy, A.P., Jain, A.K., Punch, W.F.: Combining multiple weak clusterings. In: Proceedings of IEEE International Conference on Data Mining, pp. 331–338 (2003)

35. Topchy, A.P., Jain, A.K., Punch, W.F.: A mixture model for clustering ensembles. In: Proceedings of SIAM International Conference on Data Mining, pp. 379–390 (2004)

36. Wolpert, D.H., Macready, W.G.: No free lunch theorems for search. Technical Report SFI-TR-95-02-010, Santa Fe Institute (1995)

37. Xue, H., Chen, S., Yang, Q.: Discriminatively regularized least-squares classification. Pattern Recognition 42(1), 93–104 (2009)

Author Index